Decision Support Systems for Smart City Applications

Scrivener Publishing
100 Cummings Center, Suite 541J
Beverly, MA 01915-6106

Sustainable Computing and Optimization

Series Editor: Prasenjit Chatterjee, Morteza Yazdani and Dilbagh Panchal

Scope: The objective of "Sustainable Computing and Optimization" series is to bring together the global research scholars, experts, and scientists in the research areas of sustainable computing and optimization from all over the world to share their knowledge and experiences on current research achievements in these fields. The series aims to provide a golden opportunity for global research community to share their novel research results, findings, and innovations to a wide range of readers, present globally. Data is everywhere and continuing to grow massively, which has created a huge demand for qualified experts who can uncover valuable insights from data. The series will promote sustainable computing and optimization methodologies in order to solve real life problems mainly from engineering and management systems domains. The series will mainly focus on the real life problems, which can suitably be handled through these paradigms.

Publishers at Scrivener
Martin Scrivener (martin@scrivenerpublishing.com)
Phillip Carmical (pcarmical@scrivenerpublishing.com)

Decision Support Systems for Smart City Applications

Edited by

Loveleen Gaur

Vernika Agarwal

and

Prasenjit Chatterjee

Scrivener Publishing

WILEY

This edition first published 2023 by John Wiley & Sons, Inc., 111 River Street, Hoboken, NJ 07030, USA and Scrivener Publishing LLC, 100 Cummings Center, Suite 541J, Beverly, MA 01915, USA

For more information about Scrivener publications please visit www.scrivenerpublishing.com.

Wiley Global Headquarters
111 River Street, Hoboken, NJ 07030, USA

For details of our global editorial offices, customer services, and more information about Wiley products visit us at www.wiley.com.

Library of Congress Cataloging-in-Publication Data

ISBN 978-1-119-89643-2

Cover image: Pixabay.Com
Cover design by Russell Richardson

Set in size of 11pt and Minion Pro by Manila Typesetting Company, Makati, Philippines

Printed in the USA

10 9 8 7 6 5 4 3 2 1

Dedication

The Editors would like to dedicate this book to their parents, life partners, children, students, scholars, friends, and colleagues.

Contents

Preface

The current era of urbanization requires strong strategies and innovative planning to modernize urban life. Cities play a primary role in various aspects of social and economic development worldwide, and have a huge impact on the environment. A city is a large and permanent human ecosystem that provides services and opportunities to its citizens. Promoting sustainability has been a major challenge for urban cities. The aim of sustainable cities is to ensure that cities can respond to their inhabitant's needs through sustainable solutions to socio-economic problems. The major challenge for cities worldwide is to have optimal solutions for transportation linkages, mixed land uses, and high-quality urban services with long-term positive effects on the economy.

Many cities are enhancing quality and performance of urban services by being digitalized, intelligent and smarter. Many of the new approaches related to urban services are based on harnessing technologies, including information and communication technologies (ICT), helping to create what some call "smart cities." Over the last two decades, the concept of "smart city" has become more and more popular in scientific literature and international policies. Smarter cities start from the human capital side, rather than blindly believing that ICT can automatically create a smart city. Approaches towards education and leadership in a smart city should offer environments for an entrepreneurship accessible to all citizens. The stakeholders and policymakers are exploring solutions to deliver new services in an efficient, responsive and sustainable manner for a large population. This book explores various dimensions of the possible services available for making a city smart.

Making a smart city is an emerging strategy to mitigate the problems generated by urban population growth and rapid urbanization. There is very little academic literature in this context, with much of it focused on the definition of smart cities. The growth of an intelligent decision support system presented in the literature has given rise to various computational methods. The amalgamation of these two concepts in literature is still very scarce in research. This book aims to provide a better understanding of the concept of smart cities and the application of an intelligent decision support system. Based on analysis of existing information there are eight critical factors of smart city initiatives: management and organization, technology, governance, policy context, people and communities, economy, built infrastructure, and natural environment. This book will focus on the application of the decision support system in managing these eight crucial aspects of smart cities.

Our intent in writing this book was also to provide a source that covers the stage-by-stage integration of the four key areas involving planning, physical infrastructure, ICT infrastructure and deploying the smart solutions necessary for city transformation. With this as the motivation, this book provides the application of an intelligent decision support system for effectively and efficiently managing the transformation process, which can aid various supply chain stakeholders, academic researchers and related professionals in building smart cities. Various chapters of this book are expected to support practicing managers during the implementation of smart solutions for city transformation.

Chapter 1 deals with the development of an IoT-based smart agriculture device which can be a useful product and proven game changer in Next Gen Agriculture. Aided by the internet of things, this automated smart irrigation system maintains the adequate amount of water needed by crops by monitoring the amount of moisture, temperature and humidity in the soil. Temperature and humidity data is then maintained in a database for use in crop rotation and to help farmers select the appropriate crops. It will also benefit farmers by making it possible to monitor crop irrigation from a distant location, which in turn saves farmers time and reduces their workload. Chapter 2 describes a decision support system (DSS) for smart cities. Using standard mathematical and statistical models, DSS analyzes complex data, thereby producing the required information. Since the amount of information generated/used as input always deals with big data, it requires classificational analysis. Establishing an integrated command and control center as part of the Smart City Mission of creating an intelligent traffic system is discussed and illustrated with a case study. Chapter 3 evaluates the potential interleavers for burst error control and interference mitigation, whose possibilities for use in UAV communications are yet to be investigated. Hence, the idea of employing interleavers for interference mitigation in UAV communications is envisioned and analyzed. More specifically, we explore the role of ITI in UAV communication for this purpose. The chapter also provides a decision support system used for interleaver assignment to each UAV that is applicable within the smart city framework. Chapter 4 describes the difficulties faced when enhancing the consistency and sustainability of facilities for city residents, and also presents the characteristics of the ecosystem in smart cities. In this context, a DSS based on the development of the model of the analytical hierarchical process integrated into the logic representation system has recently become increasingly relevant. The DSS is positioned to be a unique benefit as a constantly evolving tool to prepare and complete exponential growth using IT infrastructure, cloud computing, web-based apps, and everything-as-a-service (XaaS) for the development of new mathematical models, artificial intelligence and data storage. Included in the chapter are decisions, decision modeling and decision management principles, support mechanisms for decision making and collaborative systems, as well as how they can assist in smart city planning initiatives.

Next, Chapter 5 discusses how technology enhancements that include social sustainability are driving the development of smart city economies. Blockchain technology, also known as "digital ledger," is the building block for these smart cities, and the demand for it is increasing at a very rapid rate in all sectors and markets. This chapter describes the challenges which hinder the adoption of blockchain technology in smart cities and provides valuable solutions for solving them using fuzzy step-wise weight assessment ratio analysis (SWARA) and weighted aggregated sum product assessment (WASPAS). These aim to determine the weight of each challenge and its importance and provide a rank to different solutions which will help in addressing them. The study is validated by taking Indian cities into consideration. Chapter 6 discusses the implementation of a DSS in modeling a smart city supported by the internet of things (IoT), cloud services, artificial intelligence (AI), and everything-as-a-service (XaaS), which position it ahead of many other software. The primary purpose of this chapter is to identify the challenges which hold back the use of DSS in smart cities and to provide optimal solutions to them by using a multicriteria decision-making (MCDM) approach. The technique for order preference by similarity to ideal solution (TOPSIS) is a fuzzy MCDM method. The results have been validated using the case of smart cities in northern India.

Next, Chapter 7 highlights the changes that have been implemented to help people adapt to new age technology. Those in different sectors have been forced to learn new ways of working; for example, smart cities which were only in the planning and policy development stage suddenly saw "Jugaad" technology playing a role in it. Therefore, the aim of this chapter is to engender understanding of the barriers to using technology in the education sector. Since there are many stakeholders to consider in this sector, it is impossible to consider all dimensions in one study. The aim is to determine the barriers faced by teacher's training through an online teaching platform. The DEMATEL technique is a multicriteria decision-making tool to further establish the cause-and-effect between the criteria. It is necessary to remove the antecedents of the consequents and establish a better education system that leads to smart education for smart cities. Chapter 8 describes modern technology, from IoT to news. The integration of these innovations will turn industry 4.0 into industry 5.0 in the context of smart city technologies.

Next, Chapter 9 presents a framework for developing smart child safety using the internet of things; and Chapter 10 implements deep learning and transfer learning algorithms to predict soil moisture for smart city application, both with and without the effect of data augmentation. Chapter 11 presents the role of information and communication technologies (ICTs) in the sustainable development of smart cities toward futuristic communication. In the context of "smart life," Chapter 12 proposes a new framework in which the value of knowledge is defined not only by competitiveness and productivity, but also by consumer demand. It is based on a quantitative study of 300 traditional product users in a digital environment. Synergies between technological, social, and economic systems are highlighted in the contributions. The concluding Chapter 13 presents surveys and analyses of smart city projects, as well as recommendations on how to sustain them while balancing the Earth, people, and profits. Furthermore, new technical implementations are recommended that are not only enticing but also tackle real-world challenges in emerging countries like India.

The transformation of "cities" into "smart cities" is the need of the hour. Due to the growth in population and high level of urbanization, policymakers, urban developers, government officials and service providers need to develop smart cities that are self-sustaining. A smart city is a framework, predominantly composed of information and communication technologies (ICT), in which to develop, deploy, and promote sustainable practices to address the growing challenges of urbanization. A big part of this ICT framework is essentially an intelligent network of connected objects and machines that transmit data using wireless technology and the cloud. Cloud-based IoT applications receive, analyze, and manage data in real time to help municipalities, enterprises, and citizens make better decisions that improve quality of life. Citizens engage with smart city ecosystems in a variety of ways, such as using smartphones and mobile devices, as well as connected cars and homes. Pairing devices and data with a city's physical infrastructure and services can cut costs and improve sustainability. The major challenges in this framework are the application of computation techniques, including real-life data processing, connectivity issues, and application of IoT in everyday life. This book aims to address these challenges by describing how the transformation of cities into smart cities can be smoothed with the help of an intelligent decision support system.

Aimed at both academics and practitioners alike, this book was designed as a guide to developing smart cities. It focuses on the hierarchical decision-making process in the field of smart cities implementation; more specifically, it provides an essential framework in which policymakers and executives can make strategic decisions. The potential audience for the book includes researchers, industrial manufacturers, supply chain managers, stakeholders, policymakers, and other technical and administrative personnel. The book can be used as a textbook or to supplement graduate programs in smart cities, operations and management, sustainable management, and business administration.

The Editors
October 2022

Acknowledgment

The editors wish to express their warm thanks and deep appreciation to those who provided valuable input, support, constructive suggestions and assistance in editing and proofreading of this book.

The editors would like to thank all the authors for their valuable contributions in enriching scholarly content of the book.

Mere words cannot express the editors' deep gratitude to the entire editorial and production teams of Scrivener Publishing, particularly Martin Scrivener for his great support, encourangement and guidance all through the publication process. This book would not have been possible without his significant contributions.

The editors would like to sincerely thank the reviewers who kindly volunteered their time and expertise for shaping such a high-quality book on a very timely topic.

The editors wish to acknowledge the love, understanding, and support of their family members during the book's preparation.

Finally, the editors use this opportunity to thank all the readers and expect that this book will continue to inspire and guide them for their future endeavour.

The Editors

1

Techno Agri for New Cities by Smart Irrigation

Rohit Rastogi*, Sunil Kumar Prajapati, Shiv Kumar and Satyam Verma

Computer Science & Engineering Department, ABES Engineering College, Ghaziabad, U.P., India

Abstract

Agriculture is the most important source of food production. It also plays a crucial role in the gross domestic product of the country. But there are various constraints in traditional methods of agriculture. These constraints include excessive use of water during cultivation of crops, time, money, etc. In order to overcome the various constraints involved in the agriculture sector, there is a need for an evolved irrigation system. This paper aims at developing an automated smart irrigation system with the help of the Internet of things. Its aim is to maintain an adequate amount of water needed by the crop by monitoring the amount of soil moisture, temperature, and humidity in the soil. The data of temperature and humidity are maintained in the database for backup. The data are used for crop rotation and also help the farmer for the selection of appropriate crops. We can also verify the different types of soil appropriate for different crops using this model. These will also benefit the farmers as they will be able to monitor the irrigation of the crop from a distant location. It would also save the time of the farmer and reduce the labor work. The manuscript deals with the IoT-based smart agriculture device, which can be developed as a useful product and be proved as a game changer in the next-gen agriculture. In South Asian continent, farmers are economically poor, and they will be highly benefitted by this application, if developed commercially. Smart cities concept in India will be supported by this.

***Keywords*:** Arduino, soil moisture sensor, humidity and rain sensor, esp8266 wi-fi module, dht-11, smart irrigation, IoT

**Corresponding author*: rohitrastogi.shantikunj@gmail.com

Loveleen Gaur, Vernika Agarwal and Prasenjit Chatterjee (eds.) Decision Support Systems for Smart City Applications, (1–16) © 2023 Scrivener Publishing LLC

1.1 Introduction

Agriculture can be defined as a technique of cultivating the soil, growing crops and raising livestock. Agriculture is considered as the main source of food and fabrics. Cotton, wool, paper and leather are all agricultural products. Agriculture also provides wood for construction materials and other household activities. Before agriculture became an important factor people used to spend most of their lives searching for food and hunting wild animals. But around 2000 years ago, agriculture became the most important source of food and most of the Earth's population became dependent on agriculture.

Water is a basic need of every living being in this world. It plays a vital role in carrying out day to day activities in human life. Agriculture is an area where water is required in a large quantity for better growth of the crops. But due to overuse of water the ground water level is depleting very rapidly. The main reason for this problem is the population growth and its increasing demand for water requirements. Overconsumption and wastage of water is another major problem, which is leading to water crisis in this world. Water crisis may lead to economic decline and poor living conditions if we continue the current scenario of water usage [5, 12].

With the estimated growth of the world's population to 8 billion by the year 2050, the requirement for crops and food will also increase rapidly. On the other hand, the temperature is likely to increase by 4 to 5 degrees in the next few years due to global warming. Some climate models describe that there would be an increase in concentration of carbon dioxide on the crops. Therefore, climate change has the potential in affecting the productivity of agriculture. It is expected that there would be an increase in yearly dry days to about 15 extra dry days in the next few years. Which means that the dry areas would likely receive less rainfall throughout the year. This will have a direct impact on the total growth of agriculture [13].

Water is considered as the most important substance for running our life properly. Our bodies need water to function properly. According to Science humans can survive for weeks without food, but can survive only a few days without water. But we humans are depleting the fresh water sources very rapidly because we are not bothered to use the water in an efficient manner [13].

The Earth's temperature is rising due to global warming and the hotter the earth will be the more would be the demand for water. The shortage of water will lead to less production in the agricultural field and thus the water crisis will become a food crisis. The main source of freshwater is groundwater, which is decreasing very rapidly. Ground water level could be increased by using the technique of rainfall harvesting [21].

Smart irrigation devices are the components that we are using in this project which will first analyze the climatic conditions like rainfall and temperature and then the will automatically operate the process of irrigation. Devices like rainfall, temperature, and humidity and moisture sensors are able to give precise values, which are used by Arduino to carry out the automated irrigation process. These values are used to match the threshold values and then the water pump is turned ON and OFF accordingly. Thus, using these smart irrigation devices, we can reduce water wastage, as well as increase the productivity [8].

IoT abbreviates to the Internet of Things. IoT is considered as a milestone when we talk about the evolution of superior technology. IoT comes into mind when we try to automate things. IoT can be applied in various sectors such as home automation, surveillance systems, and in the agriculture sector, there is a wide range of applications of IoT. As we already know that the crops require proper care for better yielding and irrigation is the most dominating factor that affects it most. Due to irregular monsoon, cultivated plants do not grow properly and result in low production. Using advanced technology like IoT, we can overcome this problem. By planting different sensors in the field, we can record important factors like temperature, humidity of the air and soil moisture content and make decisions accordingly using microcontrollers. Irrigation will be done automatically when the moisture of soil falls. It will be more helpful in the areas where there is a lack of water supply and fewer rainfall readings. The use of IoT in an irrigation system can bring a new revolution in the agriculture sector [10, 12].

The soil moisture of the field can be figured out by various techniques, such as by using the thermogravimetric method or by using a gypsum block and tensiometer methods. These methods are old and are put back by time domain reflectometry, frequency domain reflectometry, and optical sensor technology. Soil moisture estimation based on sensors provides data, which is real time, at an affordable cost. The sensor-based irrigation has a lot of positive points over the traditional method. It collects data that are real time and can be interpreted accordingly by different smart modules. It is cost-efficient and time-saving [11, 14].

- Productivity increase,
- Less water consumption,
- Almost zero manpower consumption,

- Cost efficient,
- System have weather resistance,
- Most efficient use of water.

Drip-irrigation System (Traditional): The most efficient way of irrigation is the traditional drip irrigation system. It allows water to ooze at the plant roots, resulting in less water wastage. It also helps in the efficient utilization of fertilizer, which is absorbed by soil uniformly with steady irrigation [12, 15].

Irrigation with Timer System: The best way to reduce water wastage in irrigation is by making a schedule. An irrigation system with an automatic timer can prevent over-watering in the field and can prevent from damaging the crop due to excessive irrigation. It helps to manage the water requirement for each season. It is cost efficient and reduces wastage of water while irrigation [12].

Smart Irrigation System: It uses MATLAB along with wireless sensors and IOT. Very good for the water usage optimization and can be operated remotely. It has auto and manual mode, which are very helpful, and cloud implementation makes it highly applicable [12, 16].

Research Objective

In this fast-growing digital world, we have thrust our thinking limit and are trying to replace normal brains with an artificially created one. Using AI we can make an intelligent machine. Machine learning with deep learning, ANN, CNN, sensors can intensify the machine work, which results in the development of more superior technology. The use of AI and ML in the agriculture sector along with different sensors to capture data can bring revolution and give birth to a happy and prosperous era [6].

1.2 Literature Review

The paper by Bobby Singla and others tells about how we can effectively control the water supply in our agricultural field. Sensors that are used for this application are DTH-11 sensor and soil moisture sensor. The information is provided on farmer mobile phones using Wi-fi and Arduino. In this manuscript, the DTH-11 temperature sensor and soil moisture sensor are connected to the input pins of Arduino Uno. The analog values produced by Arduino Uno are converted to digital output by the microcontroller. The obtained values are displayed by the mobile application. The motor is switched on/off based on the value obtained from the microcontroller with the already defined threshold value. The abovementioned system is found to be efficient in reducing the cost of the farmers and optimizing their agricultural production. The maintenance required by the system is also less [1, 17].

Rawal, Shrishthi and other team has found in their paper proposes an irrigation system which maintains and decides the required soil moisture content through automatic watering. The value obtained from soil moisture sensors helps to determine the exact quantity of water needed for irrigation. The system is divided into hardware and software components. Hardware comprises systems such as sensor, Arduino-uno whereas the software consists of a webpage displaying the data from the microcontroller. The sprinkler control is achieved using a threshold value. The value obtained from the system decides whether to turn on/off the sprinkler. The reading obtained is then put forward on the farmers' website. The system uses value obtained from the microcontroller to on/off the sprinkler. This prevents the loss of the farmer and thereby avoiding crop damage [2, 18].

Nandhini and their team of researchers revealed that the proposed irrigation system helps to regulate the flow of water in the system. By using these systems, we can make effective use of water. The system uses soil and humidity sensors to find the level of moisture and humidity in the soil. The sensed values are then displayed on the screen. This system also uses various sensors, such as pH sensor, pressure sensor, DTH-11 sensor to find the sensed values from the Arduino UNO. The sensed values are then sent to be displayed on the screen of the web page application. If the value on the sensor crosses the threshold value, then the pump is turned on/off automatically. The main objective is to find the effective, user-friendly solution to the given problem. Due to readily available updates from the server, users can know about crop fields anytime [3, 19, 21].

In their experiments of agriculture, Aman Kumar and team proposed that their system is an automated irrigation system designed to save the time, power, and money of the farmer. By using these systems, we can make effective use of water. The system uses soil and humidity sensors to find the level of moisture and humidity in the soil. The sensed values are then displayed on the screen. The sensed values are then sent to be displayed on the screen of the web page application. If the value on the sensor crosses the threshold value, then the pump is turned on/off automatically. The main objective is to find the effective, user-friendly solution to the given problem. Due to readily available updates from the server, users can know about crop fields anytime [4, 20].

1.3 Components Used

Arduino System

Arduino UNO is basically a microcontroller, which has both hardware, as well as software components. Multiple sensors could be connected at a time to the Arduino board, and these sensors gave values to Arduino with the help of program codes. The board also consists of LED, which glows when our values are matched. We can run Arduino either by connecting to our computer or by using DC power supply. The main concept is to run a physical device by using software. With the help of programming, we can easily automate various devices using Arduino. An Arduino board generally consists of analog and digital pins, a USB port, a power jack as well as a reset button (as per Figure 1.1) [8].

Sensors Used in our Research Work

In this project, we are using three types of sensors in order to calculate the soil, as well as atmospheric conditions. Sensors used are soil moisture sensor, humidity, and temperature sensor, as well as rainfall sensor.

Soil Moisture Sensor

The moisture of the soil plays an important role in the irrigation of a field. The soil moisture sensor is a kind of sensor which is used to measure the content of water within the soil. Moisture of the soil is dependent on the amount of water within the soil. If the soil is dry, it will have less moisture as compared with the wet soil. The moisture sensor works by inserting it into the field and the water content in the soil is reported in the form of percentage. There are multiple uses of soil moisture sensor:

Figure 1.1 Basic Arduino Uno, microcontroller [8].

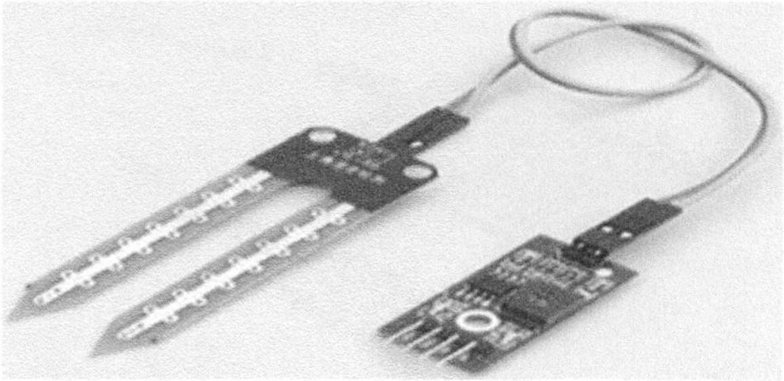

Figure 1.2 Soil moisture sensor [8].

- Agriculture
- Landscape Irrigation
- Research (as per Figure 1.2) [8].

Temperature and Humidity Sensor

Temperature and humidity sensor (DHT11) is a combined low cost sensor that gives values for both temperature, as well as humidity in the environment. It works by inserting the sensor in the Arduino board and it gives climatic conditions of the surroundings. Humidity measurements do not mean to measure humidity directly; rather they depend on the measurement of quantities such as temperature, pressure, mass, resistivity to calculate humidity. These sensors give the output as digital values, which make them easy to interface and use with microcontrollers, such as Arduino, Raspberry Pi boards (as per Figure 1.3) [8, 9].

Rainfall Sensor

The rainfall sensor is a device that is used to calculate the amount of rainfall in a particular area. This sensor is used as a water preservation device,

Figure 1.3 Temperature and humidity sensor [8].

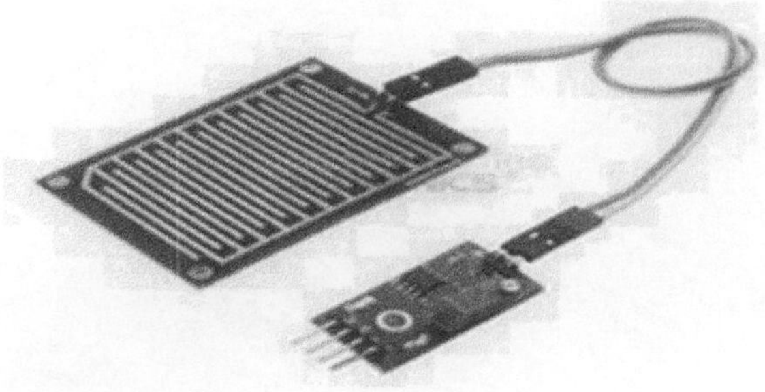

Figure 1.4 Rainfall sensor [8].

and this is connected to the irrigation system to check if there is rainfall going on and if the condition is true it shuts down the system at the time of rainfall. This sensor includes a board with nickel coated line, and it works on the resistance principle. When the rain droplets fall in the nickel coated board it gives the value of rainfall in the area. The four pins of the sensor are inserted in the Arduino while the board is kept in the field to calculate values (as per Figure 1.4) [8].

1.4 Proposed System

Various sensors, microcontrollers, the android application can be used for making an automatic irrigation system. We generally go for low-cost humidity, temperature, and soil-moisture sensors. These sensors are connected to Arduino and continuously monitor the field.

The collected data by the Arduino through sensors are transmitted to the user wirelessly so that they can control the system remotely. The smart android application compares the value received from the sensors from its database and takes the appropriate decision. The proposed system has two modes auto and manual.

When the auto mode is on, the system acts automatically without any human interruption, while with manual mode ON, the motor can be operated with just a click of the switch. The motor toggles accordingly with soil moisture value, if the value is below the threshold motor turns ON else remains in OFF state.

The sensors are joined to the Arduino Uno and the hardware communicates through a microcontroller (ESP8266) which is a wi-fi module.

All sensor values are displayed on the mobile interface so that the user has a continuous reach of the condition of the field. Programming of Arduino Uno is done in Embedded C.

We program the board that it transmits the sensor value and motor condition to the user and can also control the motor when the auto mode is engaged. The coordination of four sensors and the motor is controlled by the program fed on the board. The system continuously monitors the soil moisture content and keeps sending to the user, if the sensor gets low reading, it turns the motor on and on, reaching the requisite state it turns it off. All these functionalities are governed by a set of code fed on the board.

The user and board communicate via the wi-fi module (ESP8266). It has a quite considerable range. The threshold value will be set in the board and android application.

The moisture of the soil will be different in the winter and summer season and also the humidity and temperature. The threshold value is formulated after the consideration of different environmental and climatic conditions. The system turns the motor on automatically if the reading goes below the threshold and vice versa. The former can also operate the motor manually using the android application. Below are the block diagram and data flow diagram of the proposed model (Pl. refer thc Figure 1.5 to Figure 1.9 for the different diagram of our proposed model).

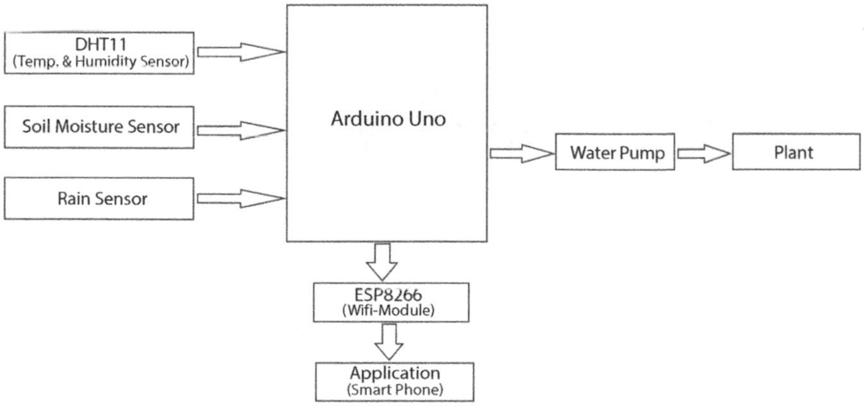

Figure 1.5 Block diagram of smart irrigation system using IoT: future prospective for agriculture.

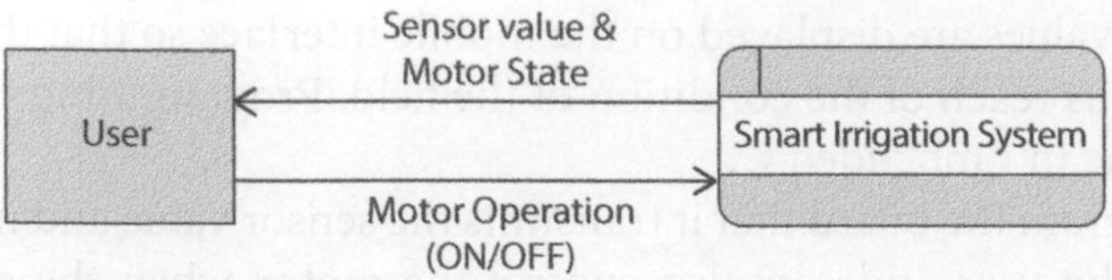

Figure 1.6 Data flow diagram of smart irrigation system using IoT: future prospective for agriculture (level 0).

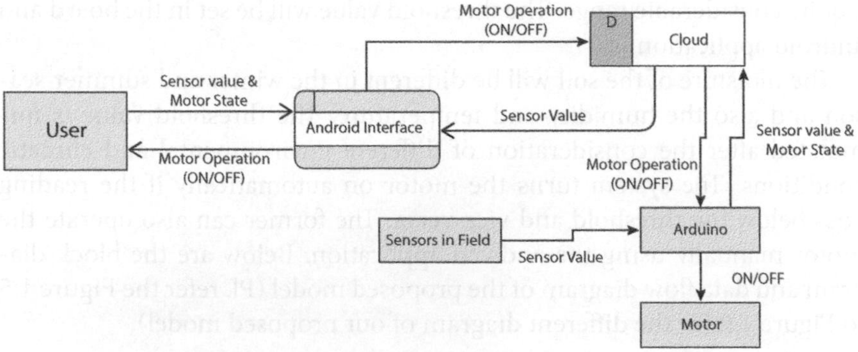

Figure 1.7 Data flow diagram of smart irrigation system using IoT: future prospective for agriculture (level 1).

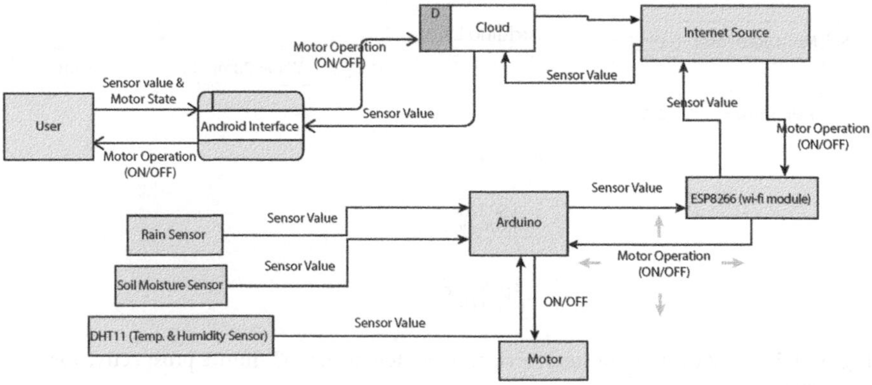

Figure 1.8 Data flow diagram of smart irrigation system using IoT: future prospective for agriculture (level 2).

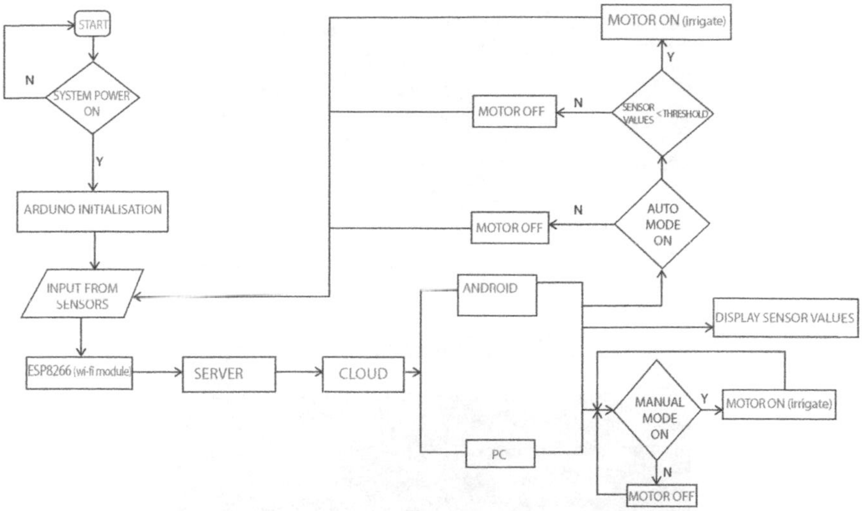

Figure 1.9 Activity diagram/flowchart of smart irrigation system using IoT: future prospective for agriculture.

1.5 Android Mobile Application for Smart Irrigation

The Android Mobile application is used to automate the activity of irrigation. The application's user interface consists of a login page where the farmer has to enter the login credentials to enter the main functioning page. The farmer first has to create his/her account in order to use the application. This is an extra security feature added so that nobody can misuse the system. There are two options on the login page, one to sign up and other to login. If the user is new, he/she first has to make his account, and then he will be able to login successfully (as per Figure 1.10).

1.5.1 Main Page

Home Page of our working App after successfully logging in from the login page, the farmer will be able to access the main page of the application. Here on this page all the values from the sensors would be displayed. If the farmer wants to manually operate according to the current values, then he can manually turn the ON/OFF button or he can use the automation button, which will automatically turn the motor ON/OFF when the Threshold values are matched (as per Figure 1.11).

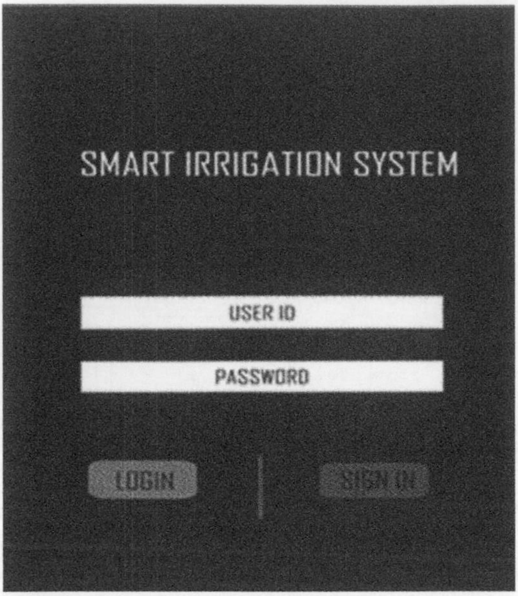

Figure 1.10 Login page of the app of smart irrigation system.

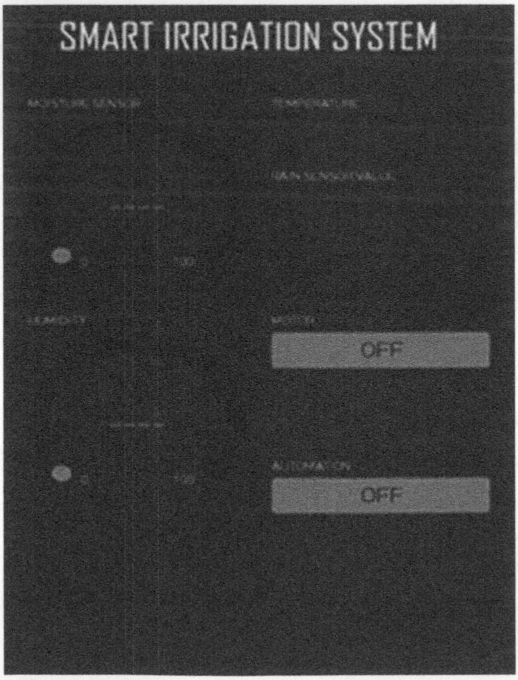

Figure 1.11 Home page of our working app.

1.5.2 Snapshot of Working Model (ICs and Working Model)

In this project, multiple sensors are connected to the Arduino board which will give their respected values when inserted in the field. We will program the Arduino in such a way that it will automate the process of the irrigation.

The user's job is to only start the motor pump or if he desires, he can switch off the motor by just one click.

On starting the motor pump, the following conditions will work:

1. The user can switch OFF the motor manually with the help of an android application.
2. The motor pump will automatically get switched OFF on reaching the threshold value of soil moisture.
3. If it is already raining in the field then the motor pump will automatically turn OFF because watering the field is not required if it is raining outside. Once the rain stops and the conditions go under the required threshold value then the motor will again turn ON. This helps in saving water resources and electricity (as per Figure 1.12).

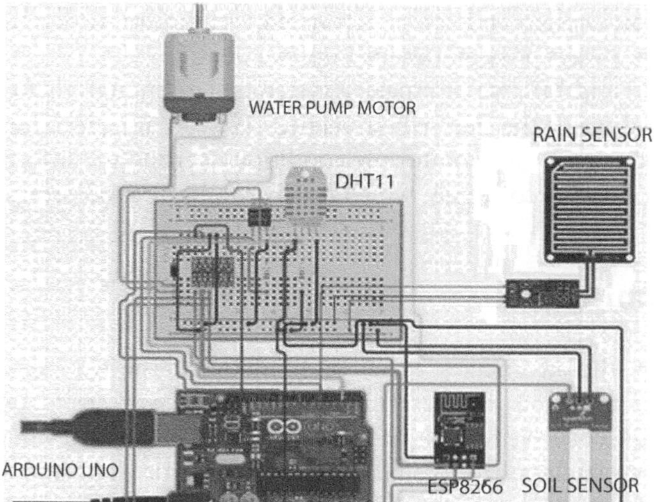

Figure 1.12 Proposed system of smart irrigation system using IoT: future prospective for agriculture.

1.6 Novelty

There are many smart irrigation projects available in the market but they are generally very expensive, and they work only for a bigger region whereas this project can also work for a small land, and it is also inexpensive and efficient as compared to other projects. With the factor of low cost, any farmer could easily buy and use it. It is also very simple in design and working is also very easy as compared to other projects so it can be used by anybody. The maintenance cost is also very low. The project aims in reducing the water wastage while working in an efficient manner.

1.7 Future Research Work

The proposed model is highly effective and is cost-efficient. It will be more effective when used along with the drip-irrigation system. There are a lot of possibilities in this model. Research can be done to make it more cost-efficient using alternate semiconductors, the interface can also be enhanced and more functionality can be added. There is a possibility of improving the server connectivity which will be very beneficial. By applying different AI and ML algorithms it can be made more advanced. Overall, the model is more than enough and can be advanced with future work and research as there are always possibilities for improvement.

Using AI and ML, we can make a system smart enough that can act on its own and water the field when needed. We can also develop a system smart enough that can monitor the condition of crops and inform the former who can act accordingly, which finally results in good production [7].

1.8 Limitations

Apart from the advantages, it has its limitations. The ICs are not much compatible with the weather and require waterproof protection which increases the cost. We have to use multiple sensors to record value as it has a small range so there are possibilities of human error while implanting the sensors. Sometimes, sensors do not work as expected which require proper attention to good and precise results. With proper knowledge and skill, this mode can be very beneficial and helpful to the agriculture sector. The limitations of the model can be reduced with good research work.

1.9 Conclusions

The abovementioned proposed model Smart Irrigation System Using IoT: Future Prospective for Agriculture is much realizable and cost-efficient for less consumption of water while irrigating the field. This system is very beneficial in the region with less rainfall, hence helps in sustainable development. Irrigation of the field can be done in a much smarter way using this model. This model will be very useful in minimizing the consumption of water while irrigation and the saved water can be used for other purposes which results in the conservation of water. This model is eco-friendly and does not harm the field in any way. With multiple sensors, we can locate the area which requires water and irrigate that area only. It requires less maintenance and is highly effective in the reduction of water consumption. With proper irrigation, the productivity increases and results in greater profit. The future work is to make the interface much more detailed and add more functionality to the system.

References

1. Singla, B., Mishra, S., Singh, A., Yadav, S., A study on smart irrigation system using IoT. *IJARIIT*, 5, 2, 1416–1418, 2019.
2. Rawal, S., IOT based smart irrigation system. *Int. J. Comput. Appl. (0975-8887)*, 159, 8, 7–11, 2017.
3. Nandhini, R., Poovizhi, S., Jose, P., Ranjitha, R., Arduino based smart irrigation system using IoT. *3rd National Conference on Intelligent Information and Computing Technologies, IICT*, pp. 1–4, 2017.
4. Kumar, A., Kumar, A., Sharma, P.K., Smart irrigation system using IoT: SIS. *Int. J. Eng. Res. Technol.*, 6, 6, 103–107, 2017.
5. Sahu, T. and Verma, A., Automated smart irrigation system using raspberry pi. *Int. J. Comput. Appl.*, 172, 6, 9–14, 2017.
6. Barkunan, S.R., Bhanumathi, V., Sethuram., J., Smart sensor for an automatic drip irrigation system for paddy cultivation. *Comput. Electr. Eng.*, 73, 9–14, 2019.
7. Math, A., Ali, L., Pruthviraj, U., Development of smart drip irrigation system using IoT. *IEEE Distributed Computing, VLSI, Electrical Circuits and Robotics, Discover 2018-Proceedings*, 2019, pp. 126–130, 2019.
8. Yamini, R. and Vishnu Chaithanya Reddy, K., Smart irrigation using IoT. *IJARIIT*, 5, 2, 467–471, 2019.
9. Jegathesh Amalraj, J., Banumathi, S., Jegathesh John, J., A study on smart irrigation systems for agriculture using IoT. *Int. J. Sci. Technol. Res.*, 08, 12, 1935–1938, 2019.

10. Dharmaraj, V. and Vijayanand, C., Artificial intelligence (AI) in agriculture. *Int. J. Curr. Microbiol. Appl. Sci.*, 07, 12, 2122–2128, 2018.
11. Gupta, L., Intwala, K., Khetwani, K., Smart irrigation system and plant disease detection. *Int. Res. J. Eng. Technol.*, 04, 03, 80–83, 2017.
12. Jegathesh Amalraj, J., Banumathi, S., Jegathesh John, J., A study On smart irrigation systems for agriculture using IoT. *Int. J. Sci. Technol. Res.*, 08, 12, 1935–1938, 2019.
13. Barkunan, S.R., Bhanumathi, V., Sethuram, J., Smart sensor for an automatic drip irrigation system for paddy cultivation. *Comput. Electr. Eng.*, 73, 9–14, 2019.
14. Bhatta, N. and Natarajan, T., Utilization of IoT and AI for agriculture. *Int. J. Adv. Technol. Eng. Explor.*, 8, 1–25, 2019.
15. Chawda, K. and Chanchal Hazra, F., Leaf based disease identification in farms, CS 229, Project Report Fall, published by www.standford.edu, 2015.
16. Dharmaraj, V. and Vijayanand, C., Artificial intelligence (AI) in agriculture. *Int. J. Curr. Microbiol. Appl. Sci.*, 07, 12, 2122–2128, 2018.
17. Garg, D., Khan, S., Alam, M., Integrative use of IoT and deep learning for agricultural applications, Springer, Cham, 2020.
18. Gupta, L., Intwala, K., Khetwani, K., Smart irrigation system and plant disease detection. *Int. Res. J. Eng. Technol.*, 04, 03, 80–83, 2017.
19. Jabal, M.F.A., Leaf features extraction and recognition approaches to classify plant. *J. Comput. Sci.*, 9, 1295–1304, 2013.
20. Kumar, A., Kumar, A., Sharma, P.K., Smart irrigation system using IoT: SIS. *Int. J. Eng. Res. Technol.*, 6, 6, 103–107, 2017.
21. Math, A., Ali, L., Pruthviraj, U., Development of smart drip irrigation system using IoT. *IEEE Distributed Computing, VLSI, Electrical Circuits and Robotics, Discover 2018-Proceedings*, 2019, pp. 126–130, 2019.

2

A Case Study of Command-and-Control Center—A DSS Perspective

Prakash B.R.[1] and Dattasmita H.V.[2*]

[1]*Department of Computer Science, Government First Grade College, Tiptur, Tumakuru, India*
[2]*Sri Siddhartha Academy of Higher Education, Tumakuru, India*

Abstract
The Decision Support System (DSS) is a vital component in smart cities. The major advances in the field of ICT have discrete impression in the field of mobility and utility services. India has latterly introduced smart cities, wherein the preliminary objective was to ensure standard and performance of various utility services through integration of information and communication technologies. Using the high level of interactivity, decision support system, including its extreme support to data access is adaptable and flexible having complete control onto the semi-structured and unstructured problems (programmed and nonprogrammed). The critical system, including the command-and-control center, completely operates on the deep learning and synthetic information processing. Despite the critical systems in the field of smart cities, the artificial intelligence applied system is extensively referred as artificial intelligence DSS (AIDSS) or intelligence DSS (IDSS). In present days, DSS and IDSS systems are greatly in use in the field of healthcare, finance, environment, security, etc.

DSS refers to a system of systems, which synchronizes with each other in the process of decision making and mostly does not intend to give a decision itself. Such systems are used to validate decision by performing sensitivity analysis on multilevel guidelines of the problem. Using the standard mathematical and statistical models, DSS analyses the complex data, thereby producing the required information. The amount of information generated/used as input is always dealt with the big data and requires classificational analysis. This paper is an attempt to discuss the intelligent transport system and command and control center

**Corresponding author*: hv.dattasmita@gmail.com

Loveleen Gaur, Vernika Agarwal and Prasenjit Chatterjee (eds.) Decision Support Systems for Smart City Applications, (17–34) © 2023 Scrivener Publishing LLC

implemented under smart cities mission and understanding the role of DSS via a case study.

Keywords: Command and control centre, decision support system, big data, critical system, fail-safe, ICMCC, safety, satellite command center

2.1 Introduction

2.1.1 Smart City

The concept of Smart City was effectuated on June 25, 2015, through the Smart Cities Mission. The main objective of the smart city is to drive economic operation and improve quality of life through all-inclusive work on social, economic, physical, and institutional pillars of the town. However, the primary goal is aimed to build a sustainable and inclusive development thereby creating the replicable models, which act as lighthouses to other aspiring cities. 100 cities are selected which were identified for the development of smart cities through a two-stage challenge process.

The implementation of the mission at the town level are going to be done by a special purpose vehicle (SPV) created for the aim. The SPV is performing the actions *viz.*, plan, appraise, approve, release funds, implement, manage, operate, monitor, and analyze the smart city development projects. Each smart city is incorporated as SPV and is headed by a full-time CEO and have nominees of central government, government, and ULB on its board. The States/ULBs is assisting by ensuring that (a) a fanatical and substantial revenue stream is formed available to the SPV so as to make it self-sustainable and will evolve its own credit worthiness for raising additional resources from the market and (b) government contribution for smart city is employed only to make infrastructure that has public benefit outcomes. The execution of projects could also be done through joint ventures, subsidiaries, public-private partnership (PPP), turnkey contracts, etc. suitably dovetailed with revenue streams.

As a centrally sponsored scheme, central government is supporting for consecutive 05 years with Rs.100 crore per annum per city. The equivalent amount on an identical basis is provided by the state/ULB. However, additional financial support or technical support will be sought by the concerned city through convergence, and funds from ULBs' own funds, innovative finance mechanisms like municipal bonds, grants under Finance Commission, other government borrowings/programs. The distinctive feature of the SPV is that major emphases is on the participation of private sector through public private partnerships (PPP).

Being a public company, the primary objective of the SPV is to create a platform for direct citizen interaction pre/post the SPV incorporations. In line with this objective, a closed assessment of citizens' expectations was captured and synthesized as smart city proposals (SCPs). All the proposals were agglomerated at the national level, contained many as 5,000 projects worth over Rs. 2,00,000 crores, of which 45% was funded through mission (SCM) grants, 21% through convergence, 21% through PPP and rest from other sources.

However, a clear definition for smart city is not found. But in the context of India, the six fundamental principles on which the concept of Smart Cities is based are,

- Community at the Core
- More from Less
- Cooperative and Competitive Federalism
- Integration, Innovation, Sustainability.
- Technology as means, not the goal.
- Convergence.

Decision Support System

Decision support systems (DSS) are interactive software-based systems that are designed to help the managerial authorities in decision making by accessing and retrieving large volumes of information, which is generated from various synchronized information systems involved in organizational business processes, city management applications, etc. A decision network (DSS) is a computerized program supporting determinations, judgments, and courses of action in a corporation/business. A DSS generally sifts through a sensitive optimizer and analyses massive amounts of knowledge, compiling comprehensive information, which is used to solve problems and in decision making. Typical information employed by a DSS includes target or projected revenue, sales figures, or past ones from different time periods, and other inventory/operation-related data.

A decision network gathers and analyses data, synthesizing it to supply comprehensive information reports. A DSS differs from a standard operations application, whose function is simply to gather data. The perfect systems analyze information and truly make decisions for the user. At the very least, they permit human users to form more informed decisions at a quicker pace.

2.1.1.1 *Characteristics of a Decision Support System*

Employing DSS objectifies that, the information is presented to the user in the easy-to-understand way. While a DSS system is very useful as the approach for programming is through user supported specifications and reports forms. For example, the synthesized information will be represented through the graphs *viz.*, bar graph.

As the data are huge, data analysis is not any longer limited bulky mainframe computers. At present, the DSS applications are not just running on desktops or laptops, they are even found on mobile devices.

This provides them the prospect to be well informed within the least times, providing the facility to make the only decisions for his or her company and customers on the go or even on the spot.

2.1.2 The Critical System

A critical system is also a system that must be highly reliable and retain this reliability as they evolve without incurring prohibitive costs. There are four kinds of critical systems: Safety Critical, Mission Critical, Business Critical and Security Critical [23].

A critical system is distinguished by the results associated with system or function failure. Functionally, critical systems are distinguished between fail-operational and fail-safe systems, according to the tolerance they exhibit to failures:

- Fail-operational — typically required to work not only in nominal conditions (expected), but also in degraded situations when some parts are not working properly. for instance, airplanes are fail-operational because they need to be ready to fly albeit some components fail.
- Fail-safe — must safely pack up just in case of single or multiple failures. Trains are fail-safe systems because stopping a train is usually sufficient to place into safe state [19].

2.1.2.1 *Safety Critical System*

A safety-critical system (SCS) or life-critical system could also be a system whose failure or malfunction may end in one (or more) of the next outcomes,

- death or serious injury to people
- loss or severe damage to equipment/property
- environmental harm.

A safety-related system (or sometimes safety-involved system) comprises everything (hardware, software, and human aspects) needed to perform one or more safety functions, during which failure would cause an enormous increase within the security risk for the people or environment involved. Safety-related systems are people that do not have full responsibility for controlling hazards like loss of life, severe injury, or severe environmental damage. The malfunction of a safety-involved system would only be that hazardous in conjunction with the failure of other systems or human error [19].

Risks of such sort are generally managed with the methods and tools of safety engineering. A safety-critical system is supposed to lose but one life per billion (109) hours of operation. Typical design methods include probabilistic risk assessment, how that mixes Failure Mode and Effects Analysis (FMEA) with Fault Tree Analysis. Safety-critical systems are increasingly computer-based.

The key attributes that assist and manage the Safety-Critical Systems are:

- Fail-operational systems.
- Fail-soft systems.
- Fail-safe systems.
- Fail-secure systems.
- Fail-Passive systems.
- Fault-tolerant systems.

2.2 Command and Control Center—A Critical System

Introduction on Command Center

The control and command center are often designated as Integrated City Management and command and center (ICMCC). The command center conceptualized in late 19th Century [22] where the first objective was to watch and command through a centralized center called "War Room." Entire base operations were managed through the room.

There are different types of command center when categorized operationally. They are,

- Data center management
- Business application management

- Civil management
- Emergency (crisis) management

Integrated City internal control Centre (ICMCC) involves leveraging of the knowledge provided by various departments and providing a comprehensive response mechanism for the day-to-day challenges across the town. ICMCC may be a fully integrated, web-enabled solution that gives seamless incident-response management, collaboration, and geo-spatial display.

Case Study

Introduction

The ICMCC, Tumakuru, makes it easier to examine and control the identified locations in a completely automated environment for better service monitoring, regulation, and enforcement. The ICMCC is accessible by operators and anxious authorized entities with necessary role-based authentication credentials. Activities at the ICMCC will comprise of monitoring services, incident management and response as per the quality Operating Procedures with defined escalation procedures. Integrated Dashboard for entire project – ICMCC features a centralized dashboard for entire smart city project for the reporting and viewing of all the project components and key performance indicators of systems like CCTV Camera closed-circuit television, Intelligent Signaling, Smart Water Quality Monitoring, Environmental Monitoring etc. through one interface [12].

ICMCC will act because the centralized monitoring and decision-making hub for managing Smart Applications and related Infrastructure project activities. ICMCC may be a pan city infrastructure implemented to involve most digital assets onto a standard platform ensuring confluence of knowledge from multiple sources, applications, sensors, receivers, objects, and other people. this may provide a price effective, top quality reliable data to the administrations and departments for the higher functioning of those departments. The ICMCC is structured across two layers wherein Master System Integrator at the State Level are going to be hosting and managing the centralized Data Center and Data Recovery policies alongside of the smart city implementation by the Local System Integrator at the respective smart cities [16].

Technical Architecture

The Integrated City Management Control Centre (ICMCC) (Figure 2.1) is the central hub for all online data and information on smart services. It is a fully functional system that grants all operation privileges to the users who

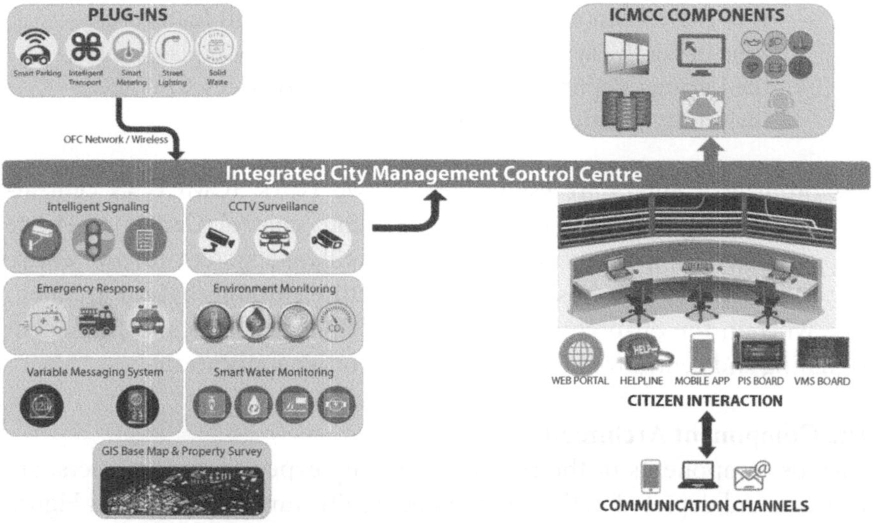

Figure 2.1 Illustrative architecture of integrated view of pan smart city components at ICMCC (Ref. Tumakuru Command and Control Centre).

have been assigned to it. This is having integration with various components with data feed view and sharing. It is also enabled with management console to perform all the operations. Full operations will be performed on the following components.

1. [18]
    1. Intelligent Signaling
    2. CCTV Surveillance System
    3. Variable Messaging System
    4. Emergency Response System
    5. Environmental Monitoring
    6. Solid Waste Management
    7. Smart Parking
    8. Street Light Control System

The benefits of the ICMCC are measured from two perspectives:

1. In times of disaster – The solution will improve the response time, coordination amongst various agencies and faster restoration of services post disaster.
2. In day-to-day non-emergency scenarios – The solution will improve the response time of Municipal Corporation

towards citizen complaints, which in turn will lead to citizens using the system more frequently. This will also enable the various departments to offer better services to citizens including, utilities, security, traffic, etc. [20].

Command Centers are established in two types, ICMCC as centralized controlling Infrastructure and a Satellite control Center at Police Department to implicitly monitor the surveillance system for the entire city. Whereas the data hosting is infrastructure through hybrid architecture wherein the on-premises for the CCTV Surveillance and Intelligent Signaling and centralized data center for all other applications.

The Component Architecture

Various components of the project, including expected system users, are as below and depicted in the component architecture as depicted in Figure 2.2.

- Street IT Infrastructure Layer
- Network Layer
- Data Center Layer
- Application Layer
- Integration Layer

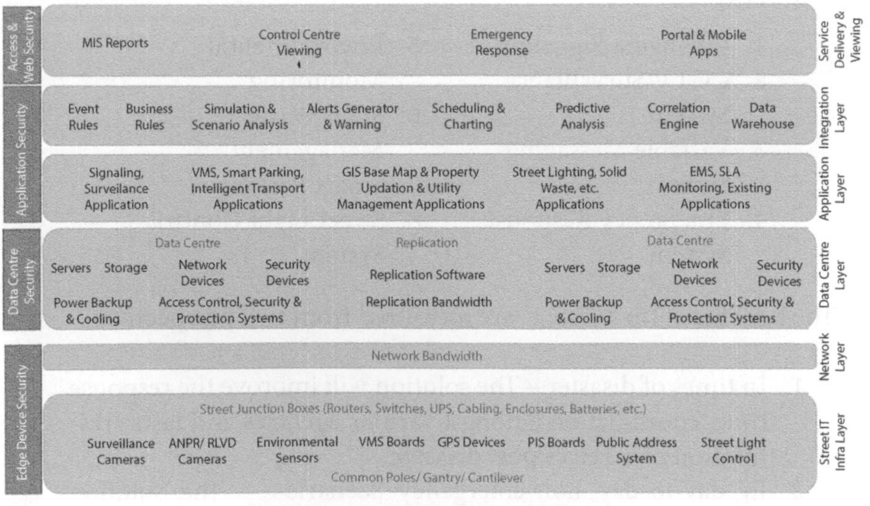

Figure 2.2 Illustrative architecture of integrated view of pan smart city components at ICMCC (Ref. Tumakuru Command and Control Centre) [20].

- Command and Communication Center Layer
- Security Layer

This component architecture is indicative in nature and is given to bring clarity on the overall scope of project and its intended use. The architecture of the complete network of smart elements is as follows.

- Street IT infrastructure layer—The sensor layer will help the city administration gather information about the ambient city conditions or capture information from the edge level devices like GPS devices, cameras, etc. Authority is expected to deploy multiple environmental sensors across the city, to measure ambient conditions such as light intensity, temperature, water level (for chronic flood spots), air pollution, noise pollution and humidity, etc. The output field devices layer will contain display devices or bi-directional (input and output) devices connected to the network, which will be used by citizens to consume—and for administrators to provide—actionable information. Such field devices include digital messaging boards, environmental data displays, PA systems, surveillance cameras among others [4].
- Network layer—the secured network layer will serve as the backbone for the project and provide connectivity to gather data from sensors and communicate messages to display devices and actuators. The SI will size the bandwidth required for the overall solution, and supply and install the edge devices to utilize the network.
- Data center layer—the data center layer will house centralized computing power required to store, process and analyze the data to decipher actionable information. This layer includes servers, storage, ancillary network equipment elements, security devices and corresponding management tools.
- Application layer—the applications layer will include applications that interface and control the street infrastructure, enterprise management system to monitor and manage all IT infrastructure and street infrastructure deployed in the city, and surveillance applications.
- Integration layer—while aspects of ambient conditions within the city will be gathered through various sensors deployed as a part of the solution, some city specific data will

come from other government and non-government agencies. Via the integration layer—that data will be exchanged to and from the under lying architecture components and other data from system developed by government (such as police department, meteorological department, streetlights department, water department, irrigation department, transport organizations within Tumakuru, etc.) and non-government agencies.

- Command and communication center layer—the command center will enable citizens and administrators alike to get a holistic view of city conditions and make informed decisions.
- Security layer—as ambient conditions, actuators and display devices are now connected through a network, security of the entire system becomes of paramount significance and the system integrator will provide: infrastructure and network security, identity and access management, and application security [20].

ICMCC as Decision Support System

Key Performance Indicators for ICMCC as decision support system, a system that has,

1. To facilitate an integrated view of all the initiatives undertaken in the city with the focus to serve as a decision support engine for city administrators for day-to-day operations or during exigency situations.
2. A fully integrated, web-based or client-server integrated with web-based access/services, solution that provides seamless incident—response management, collaboration, and geo-spatial display.
3. To make it easier to examine and operate the specified field locations in a completely automated environment for better service monitoring, regulation, and enforcement.
4. Various smart elements can use the data and intelligence gathered from operations of other elements so that civic services are delivered lot more efficiently and in an informed fashion.
5. A 24*7 City Surveillance System for effective management of the city.

6. To leverage state of the art technology to effectively manage Road Traffic.
7. To integrate with various utility systems such as intelligent transport, intelligent signaling sewerage/drainage system, disaster mgmt. system, etc.
8. To support business intelligence like multiple visuals in a single screen for intelligent decisions.
9. To serve as a decision support engine for city administrators in day-to-day operations or during emergency situations.
10. To enable Edge computing capability so that local event processing, transformation, analytics, decision & Control are done at edge devices and only business relevant events should go at data center platform.
    - A decision support tool for assessing strategies to minimize congestion, delays, and emergency response time to events via simulation and planning tools with real time traffic data fusion and control of traffic Signaling infrastructure on ground.
    - To forecast the traffic state with respect to current incidents and traffic management strategies (e.g., traffic signal control or variable message signs), improving the decision-making capabilities of the operators even before problems occur.
    - GIS as a Decision support system to prioritize actions.

Workflow, Processes, and Operations

Processes

The ICMCC is the centralized component that holds all the departments together and gives the decision makers a complete picture of the issues currently being faced by different stakeholders. The process and workflow executed within the ICMCC are:

- Definition of role-based access mechanism
- Definition, logging, and periodic review of standard operating procedures
- System health monitoring for each of the smart city components integrated with ICMCC.
- Periodic review of risks identified under monitoring and mitigation plan.

- System backup and archival process
- Audits and (Re)certifications.

Operations

- Normal Operations: Various members of the Operations team shall coordinate their operations and communicate information about the equipment/facilities under their supervision through voice and data communications technologies under normal operating conditions to facilitate a safe and secure arrangement throughout the City.
- Degraded Operation: Degraded modes of operation occur when certain systems fail to meet the levels of service expectation. In such scenario the applicable Standard Operating Procedure (SOP) would be followed.

Emergency Operation: Emergency operations are enforced in case of an unforeseen or abnormal situation when it is not possible to carry on the services. Use Cases

Name of Use Case: Motion tracking using surveillance cameras to record video feeds and take snapshots of the moving objects and send notifications.

- Purpose: Motion tracking using surveillance cameras (PTZ) (setting up a tour of presets)
- Goals: Track objects in motion and auto focus the object on motion in a specific area of camera vision during specified hours and record videos and take snapshots. The camera will receive and send alerts to take specific actions like sound alarm or alerts are used to use other cameras to record videos and take snapshots.
- Actors: CCTV surveillance operator, surveillance officer at SP office
- Smart Applications: video management server, video analytics
- Description: The network camera is configured through VMS to find objects in motion and auto focus to record videos and take snapshots. The VMS is configured to send alerts to command control center which can be used to trigger events to send alerts to other cameras to record video and take snap shots of the object and send notifications.

- Preconditions: network cameras attached to video management server is configured to detect objects in motion and follow them.
- Post-conditions: Identify incident in the command control center and run relevant SOP or computer aided dispatch to take control of the situations.
- Trigger: Objects in motion identified by network camera to send alerts
- Constraints: Camera cannot figure out if two objects are crossing on focused motion area specified Main flow:
  - VMS attached to specific network camera is configured to detect objects in motion with specific sensitivity at a specific area of camera vision.
  - The VMS can be scheduled to run the program at specific intervals.
  - The VMS runs the program continuously to detect the objects in motion.
  - When an object is seen moving it will send alert to VMS server which can be used to take various actions like record video, take snapshots and send alerts.
  - The videos and snapshots are saved for further reference to incidents.
  - Command control officer can use PAS system to make announce to control the situation.
  - Command control officer can use SOP to dispatch relevant team to the location, as necessary.

Name of Use Case: Making way to Emergency vehicles/Ambulances.

- Purpose: Emergency Scenario – Ambulance to be moved quickly to the hospital
- Goals: Dispatch notifications to emergency Response Team for rescue with a single operation (SOP) to provide timely help to move the patient from the location to a nearest relevant hospital
- Actors: Citizens, Traffic Police, Nodal officer, Ambulance driver
- Smart Applications: Smart City Control Center Integration platform, Mobile App for Ambulance driver, Intelligent

Traffic Management Application, GIS locations of Emergency Vehicles.

- Description: The citizens or any personal can call Command control center for help using phone, App Emergency box. This use case help polices personal with the route information of ambulance from the point of dispatch to the patient and hospital. Traffic police will control the traffic signals based on the destined route to move patient to the hospital and log incident data in the command control center.
  - ICMCC helpdesk receives alert.
  - Supervisor will run the SOP to information ambulance and contact driver.
  - The accumulated consumption of the current day, week, month or year, and the corresponding cost
  - How the total or seasonal consumption is allocated throughout the day (night flow consumption, day flow consumption) and the corresponding cost.
  - How is the total consumption allocated throughout the year (winter and summer consumption) and the corresponding cost?
  - The system provides options for the selection of time interval, through a calendar, and the time resolution of the presented information. The user can also obtain water consumption as volume (e.g., liters), per capita volume (e.g., liters per person), cost, per capita cost, etc.
  - Log a complaint to the nodal office related to issues with water/meter/mobile application/web site, etc.
- Preconditions: availability of ambulance, location of ambulance, location of hospitals on GIS map, operator aware of type of mishap to dispatch to relevant hospital or as per patient attendant request, police personals deployed at each location in the city and their contact information, communication channel, command control traffic controller.
- Postconditions: Information to relatives or care takers, FIR for mishaps, update incident details in incident management application, death information and certificate
- Trigger: police, citizens or relatives call command control center for help.
- Constraints: availability of real time data of traffic, police and available hospital details, details of health issues

- Main flow:
  - Incident details received from the location to control center.
  - Help desk analyzes the situation and figure out SOP.
  - Operator to follow SOP defined as per trigger and co-ordinate with ambulance, Traffic police, City police through SOP.
  - Incident data are created automatically and/or manually regarding the emergency while running the SOP.
  - Dispatch an ambulance, police, display information on VMS regarding emergency to citizens and request to make way for ambulance.
  - All the respective team's co-ordinate and help ambulance pick the patient.
  - Emergency operator analyzes and share faster route for the ambulance to reach to the patient and back to the hospital.
  - Change switching cycles of traffic signals.
  - Incident is updated with all relevant details.

2.3 Conclusion

Smart Key to Safe City, Under Pan City, a comprehensive package implemented to safeguard the city with the implication of the security, safety, quality and technology. An advanced monitoring shall be housed to foresee and uproot the loopholes across the city and thrives to protect the public interest thereon [24]. The different components are monitored, and quality controlled in the control center wherein commanded to induce the safety regulations and integrity across city. The major objectives of command center being, safety improvement, real time information, event tracking & response, and fast access to stored information, create a platform for sharing traffic information across the city and provide integrated services and information ensuring efficiency and easy accessibility. The decision support system plays a vital role in critical systems. The AI-based learning capabilities in the critical system has enabled for the more intuitive applications. The ICMCC model is very reactive and highly intuitive and found to be an effective solution through DSS. The algorithms henceforth used through DSS has enormously increased the revenue of the city through the processed workflows [17].

DSS is the basic workflow that redefine the concept of analysis and intelligent solutions for such critical system. With such interventions, DSS is proven to be an efficient solution in case of ICMCC critical system [20].

References

1. Alyaev, S. *et al.*, *An Interactive Decision Support System for Geosteering Operations*, SPE BERG, April 18 2018. Internet, June 01 2021.
2. Ben-Zvi, T., Measuring the perceived effectiveness of decision support systems and their impact on performance. *Decis. Support Syst.*, 54, 1, 248–256, 2012. doi:https://doi.org/10.1016/j.dss.2012.05.033.
3. Brynielsson, J., Using AI and games for decision support in command and control. *Decis. Support Syst.*, 43, 4, 1454–1463, 2007. doi:https://doi.org/10.1016/j.dss.2006.06.012.
4. Center, Command, Data Projections Inc. [Website], 2021. https://www.dataprojections.com/dp-control-room-solutions/command-center/.
5. Kim, D.J., Ferrin, D.L., Rao, H.R., *Decision Support Systems*, 44, 544–564, January 2008. Internet. May 30 2021. https://www.sciencedirect.com/science/article/abs/pii/S0167923607001005.
6. Fogli, D. and Guida, G., Knowledge-centered design of decision support systems for emergency management. *Decis. Support Syst.*, 55, 1, 336–347, 2013. doi:https://doi.org/10.1016/j.dss.2013.01.022.
7. Gregoriades, A. and Sutcliffe, A., A socio-technical approach to business process simulation. *Decis. Support Syst.*, 45, 4, 1017–1030, 2008. doi:https://doi.org/10.1016/j.dss.2008.04.003.
8. GSMA, March 20 2021. GSMA, https://www.gsma.com/iot/smart-cities-resources/smart-cities-safety/.
9. Irannezhad, E., Prato, C.G., Hickman, M., An intelligent decision support system prototype for hinterland port logistics. *Decis. Support Syst.*, 130, 113227, 2020. doi:https://doi.org/10.1016/j.dss.2019.113227.
10. Kamsu-Foguem, B., Tchuente-Foguem, G., Allart, L. *et al.*, User-centered visual analysis using a hybrid reasoning architecture for intensive care units. *Decis. Support Syst.*, 54, 1, 496–509, 2012. doi: https://doi.org/10.1016/j.dss.2012.06.009.
11. Lacinak, M. and Ristvej, J., Science Direct. *Procedia Engineering*, 192, 522–527, 2017. https://www.sciencedirect.com/science/article/pii/S187770581732636X.
12. PMC. Command and Control Center, Project Report, Eprocurement, Tumakuru, 2019-2020.
13. Power, D.J., 1, Data-Intensive Decision Support, Volume 1, D.J. Power, October 21 1997. http://dssresources.com/papers/whatisadss.

14. Arinze, B., A contingency model of DSS development methodology. *J. Manag. Inf. Syst.*, 8, 1, 149–166, 1991.
15. Segal, T., *Decision Support System—DSS*, M. James (Ed.), Ministry of Housing and Urban Affairs, Government of India, New Delhi, March 19 2021, https://www.investopedia.com/terms/d/decision-support-system.asp.
16. Shukla, C., *Integrated Smart City Command and Control Centre*, Bhopal Smart City Development Corporation Limited, Ministry of Housing and Urban Affairs, Government of India, New Delhi, 2017, https://smartnet.niua.org/content/9cc425c8-87cd-4908-9138-537599b270ec.
17. Simply Learn, Decision support system (DSS), Tutorials Point, April 2021. https://www.tutorialspoint.com/management_information_system/decision_support_system.htm.
18. Smart Cities mission, Ministry of Housing and Urban Affairs, Government of India, New Delhi, March 2021. https://smartcities.gov.in/.
19. Tamimi, N. *et al.*, *An Artificial Intelligence Decision Support System for Unconventional Field Development Design*, July 22 2019, Website. June 1 2021, https://onepetro.org/URTECONF/proceedings-abstract/19URTC/1-19URTC/D013S012R002/160161.
20. Team, Technical, Technical Report, Smart City, Tumakuru, 2018.
21. Smart City, Thiruvananthapuram, 2021, https://www.smartcitytvm.in/projects/centralised-command-control-centre/.
22. Wikipedia, Command center, Wikipedia, The Free Encyclopaedia, 2021. https://en.wikipedia.org/wiki/Command_center.
23. Critical system, Wikipedia, The Free Encyclopaedia, March 19 2021. https://en.wikipedia.org/wiki/Critical_system.
24. Safety critical system, Wikipedia, The Free Encyclopaedia, March 20 2021. https://en.wikipedia.org/wiki/Safety-critical_system.

3

Inverse Tree Interleavers in UAV Communications for Interference Mitigation

Manish Yadav[1]*, Prateek Raj Gautam[2] and Pramod Kumar Singhal[3]

[1]Department of ECE, Amity University, Noida, U.P., India
[2]Department of ECE, MNNIT, Allahabad, U.P., India
[3]Department of Electronics, MITS, Gwalior, M.P., India

Abstract
Fifth-generation (5G) communications integrate the role of unmanned aerial vehicles (UAVs) to achieve certain objectives. Interference among 5G users can be a serious problem for which a simple but effective solution is essentially required. In modern-day communication theory, inverse tree interleavers (ITI) are emerging as one of the potential interleavers for burst error control and interference mitigation. However, the possibilities of employing interleavers for interference mitigation in UAV communications are yet to be investigated. Hence, the idea of employing interleavers for interference mitigation in UAV communications is envisioned and analyzed. More specifically, we explore the role of ITI in UAV communication for this purpose. The chapter also provides a decision support system used for interleaver assignment to each UAV, which is applicable within the smart city framework.

Keywords: 5G, inverse tree interleaver ITI, FLRITI UAV, drone, interference mitigation

3.1 Introduction

In the fifth-generation (5G) of telecommunication, unmanned aerial vehicles (UAVs) supported communication is an important use case. The UAV-based communication is popularly known as drone communication.

**Corresponding author*: ymaniish@gmail.com

Loveleen Gaur, Vernika Agarwal and Prasenjit Chatterjee (eds.) Decision Support Systems for Smart City Applications, (35–52) © 2023 Scrivener Publishing LLC

A UAV can play any role in this communication, i.e., role of user equipment (UE), a base-station (BS) or a relay device [1–4]. The mobility, ease of deployment, quick upgrade and cost-effectiveness are the important benefits this mode of communication. The UAVs can be deployed in multiple applications, including civil, defense, Internet-of-Things (IoT), etc. [5–8].

A smart city is defined as a geographical area that employs multiple electronic processes and sensors to gather important data. The various meaningful insights obtained from this data are used to manage resources and services in an optimized manner for enhancing the functions across the city [9, 10]. The UAV-based communication can be a great support in smart cities. An improved local connectivity can be assured through this mode [11–13]. Even in emergency cases, smart cities rely on UAV-based communication. Therefore, both power saving and interference mitigation are two important aspects in UAV communications [12–14].

ITIs are one of the types of derived interleavers that are generated from modification of the tree-based interleavers. It introduces flipping of left-most and right-most bit indices of a mother seed interleaver to achieve greater potentiality [15, 16].

In this chapter, we aim to throretically integrate the three different concepts namely ITI scheme, a smart DSS and UAV communications together to obtain a well-connected smat city framework. This approach is also helpful in mitigating interference among UAVs and saving its power.

The remaining part of the chapter is arranged in different sections including background, problem, motivation, ITI-based interference mitigation, its role in smart city framework, etc. In the next section, brief backgrounds of the UAV communications, ITI, smart city framework have been covered.

3.2 Background

A decision support system (DSS) is a basic system that helps in making certain decisions in the environment where it is deployed. A smart DSS employs a set of algorithms which are based on either artificial intelligence and/or non-artificial intelligence [9, 17]. These algorithms are enablers in decision making process and reduce the manual involvement in decision making process once relevant rules are set. However, the DSS algorithms can be updated, when required. The DSS for smart cities have been explored extensively in the existing literature specifically published in the last few years [17–19]. Virgil described the role of the DSS for enhancing the existing tools involve in the process of planning and completing complicated smart cities projects [9]. Hamaguchi *et al.* gave an overview of possible roles of telecommunication for the infrastructure of smart cities [20]. Jawhar *et al.* recognized the networking protocols which can be employed to support the diversified traffic flows [21]. Maitakov *et al.* focused on reduction of costs of developing and implementing DSS in a specific subject area [22]. Ahmad *et al.* reviewed the DSS from a smart city perspective [10].

Power saving in unmanned has been identified as one of the major challenges in recent years. An energy-efficient UAV system mechanism must be placed to save its power [12–14]. More specifically, the aim of developing an energy-efficient UAV system mechanism to save its power often concentrates on saving its operational power. There are easily available literatures that cover the comprehensive review on this topic [13, 14]. Apart from power saving, interference among different UAVs is another challenge. Different solutions have been provided to overcome interference in UAV communications [23, 24]. Uragun, Amoiralis *et al.*, and Boukoberine *et al.* reviewed various optimization techniques for energy efficiency, power supply, and energy management in UAVs and their different aspects [12–14].

Fouda *et al.* suggested an algorithm for interference management that maximizes the total sum-rate of the integrated access backhaul network in UAV communications [24]. Yujie *et al.* disclosed an approach of interference mitigation in heterogeneous cloud radio access network (HCRAN) with the help of inter-tier cooperation [25]. Yatong *et al.* provided intercell interference (ICI) control mechanism in which learning process was fully decentralized [26].

Wu *et al.* and Ping *et al.* researched over IDMA system [27, 28]. Vázquez-Araújo *et al.* explored the IDMA system with low-rate layered low-density generator matrix codes [29]. Yadav *et al.* explored the same system with inverse tree interleavers respectively [15]. Modern Fisher-Yates algorithm was also implemented within IDMA system framework in Yadav *et al.* [30] and Shokeen *et al.* [31]. Detailed theory of interleavers and interleaving approaches are also available in Yadav [32, 33].

Most recently, a new form of interleaver known as inverse-tree interleavers (ITIs) is proposed and elaborated in Yadav [15, 16, 19]. These are described as efficient interleaver that requires low storage memory and involves low complexity. In the existing literature, ITIs have been named differently such as flip left-right approach based inverse-tree interleavers (FLRITIs) or invert-tree based interleavers (ITBIs) [15, 16, 30–33]. In the last few years, these interleavers have gained popularity because of their

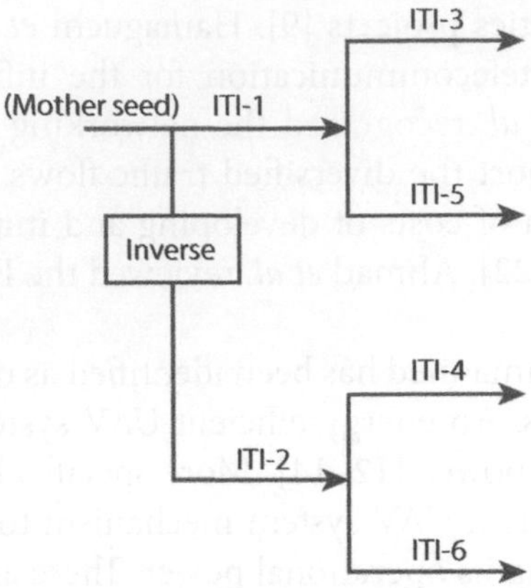

Figure 3.1 Basic ITI (or FLRITI) concept.

extremely diversified applications in domain of fifth-generation (5G) vehicular communications, internet of things (IoT), cellular IoT (CIoT), etc. [34]. Similar to ITIs, another form of interleaver, i.e., multiplicative interleaving with tree algorithm (MITA) is defined in [35]. In fact, the ITIs have been well compared with other interleavers for advanced communication systems in some available literatures on ITIs [15, 16, 19, 33]. Blockchain framework for drone communications is explored in [36].

Figure 3.1 represents the basic diagram of ITI/FLRITI generation concepts for six-user case. In this particular case, a mother interleaver seed is shown as ITI-1 and its inverse is used as ITI-2. The ITIs for other users, i.e., ITI-3, ITI-4, ITI-5 and ITI-6 are obtained as per the process explained in Yadav *et al.* [15, 16].

The bit-error-rate (BER) performance is used as a standard measure of performance for a communication system as indicates the chances of occurrence of error in a real-time scenario. For communication systems, like IDMA, the BER performances considering coded and uncoded scenarios and ITIs presence have been simulated and presented in various recent research works [37–39].

3.3 The Problem

When multiple UAVs are in close vicinity of one another, their transmitting signal may interfere, i.e., the transmitting power of atleast one UAV is strong enough to affect other UAVs boundary region. Hence, the issue of interference comes into picture when these UAVs have overlapping coverage region. The UAVs in any of its role, i.e., either working as base station or as a UE or as a relay node, often move from one position to another.

Therefore, it is very common that they arrive in the vicinity (or coverage region) of one another.

One solution for the interference mitigation could be the adjustment of height of the UAVs when their coverage regions overlap result severe interference. The problem with this conventional solution is the power limitation. The UAVs have limited power and this is why it is not desirable to waste the considerable amount of UAVs' power on their movement to mitigate interference alone. In fact, the UAVs are highly expected to spend major part of their power to enable communication and transmitting data between the end-users.

Due to the abovementioned reasons, an interference mitigation method for UAVs is required that can overcome these challenge and efficiently utilize the UAVs power.

3.4 Motivation

Smart cities envisioned a network connected with everything that exists within the city by some digital means. To facilitate this, the places where direct connectivity can be a challenge, UAVs are deployed. This is because satellite communication is comparatively a costly affair for normal user and normal communication. However, UAVs can also be equally helpful in emergency communications.

In wireless communication, it is now a well-establish fact that the interleavers have inherent virtue of mitigating the interference. This is achieved through the division and patterned (or random) distribution of long data streams over multiple blocks. The interleavers also overcome the challenge of forward error correction (FEC) capability limitation [15, 19, 40–42]. These facts are the main motivation factor behind the use of interleavers in the proposed solution for UAV communications.

In the next section, the detail aspects of the concepts and theory that have been envisioned through this chapter are provided.

3.5 Interference Mitigation Using ITI

This chapter includes the following novel aspects, which make it differ from the existing prior arts available on UAVs, interleavers and interference mitigation.

- Interference mitigation using interleavers in UAV communications—Interference mitigation using interleavers in traditional mode of communication is known. This chapter attempts to carry forward the same legacy in UAV communications.
- ITI/FLRITI in UAV communication—The idea of employing ITI as an interleaver in UAV communications for interference mitigation is also envisioned in this chapter.
- Decision support system for interleaver allocation—The concept of using a decision support system in UAV communications primarily for ITI assignment to each UAV is also provided in this chapter. This concept is equally applicable within smart city framework too.

3.6 Interleavers for Interference Mitigation in UAV Communications

Interleavers are the great mean to spread the data streams. Whenever, there are other UAVs present in the vicinity of a UAV, the problem of interference may arise due to their overlapped coverage region. To avoid this, interleavers can be employed. The data bits from each UAV are dispersed using a certain pattern and then transmitted. This pattern is a unique pattern that cannot be identical for any two UAVs. By this way, it is possible to overcome the problem of interference between two or more UAVs located in close vicinity and/or have an overlapped coverage region.

Figure 3.2(a) depicts the basic idea of employing interleavers for interference mitigation in UAV communications. On detecting the interference, a UAV approach to its controlling base station for the interleaver assignment and the base station assigns an interleaver (i.e. a unique pattern for data stream shuffling) to the UAV. Figure 3.2(b) illustrates assignment of different interleavers to the two different UAVs through minimum signaling message exchange. Here, interleavers ensure that UAVs need not to adjust their height to achieve interference mitigation, and therefore, power saving is obtained.

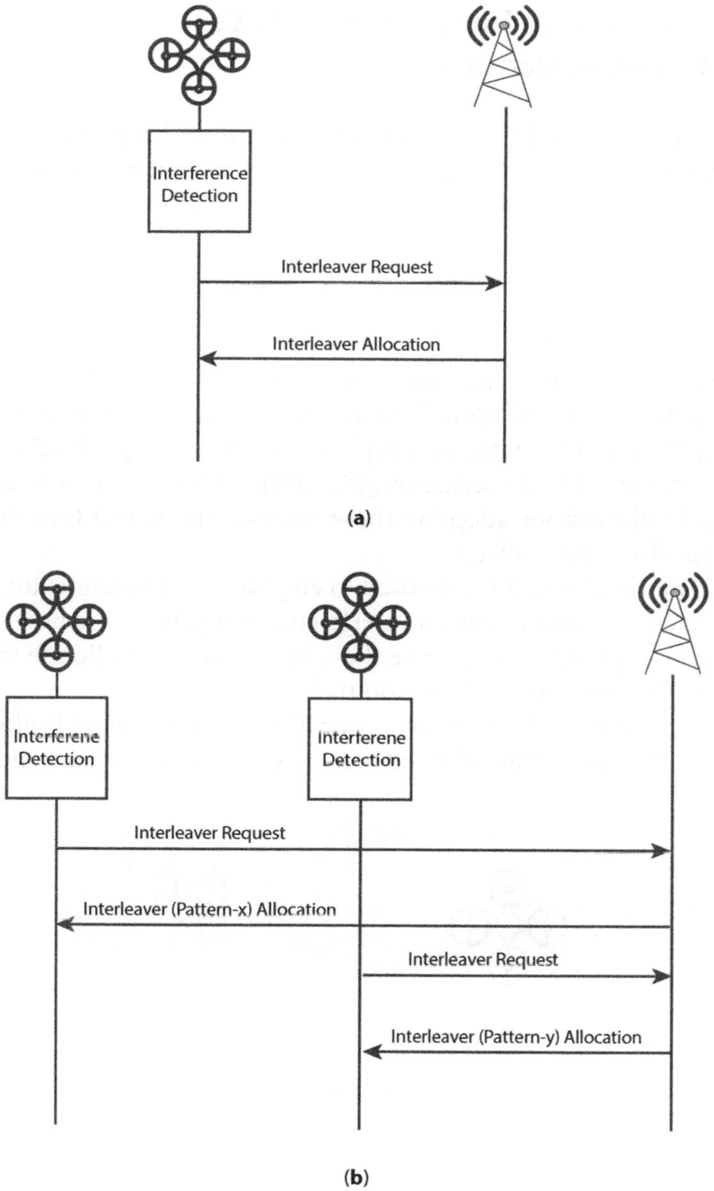

Figure 3.2 (a) Interleaver allocation in UAV communication. (b) Interleaver allocation in UAV communication with unique pattern for each UAV.

3.7 Inverse Tree Interleavers in UAV Communications

In the last few years, ITI (i.e., FLRITI) has already been emerged as an outstanding interleavers for various types of communications due to their immense capabilities, robust performance and extremely virtuous properties such as lower complexity, simplicity, i.e., simple algorithm for generation and lower memory need for storage. Initially, these interleavers were developed for multiple types of interference mitigation in interleaver-based systems i.e. IDMA, SCFDMA-IDMA, OFDM-IDMA, etc. Then, they have been explored for other advanced communication applications, e.g., blockchain, cellular IoT, optical IDMA and power line communications, etc. [34, 35, 43]. They were also explored for 5G multiple Radio Access Technology (multi-RAT) scenarios [44]. Over the time, various issues that came as bottleneck for adopting these interleavers in real-time systems have been also been resolved.

Based on the above, it is proposed to employ ITI as potential interleavers in UAV communications for interference mitigation. The base station runs the ITI algorithm to generate these interleavers and allocate them to different UAVs in real-time based on their request.

Figure 3.3 shows the simplistic signaling diagram for ITI allocation in UAV communication. This involves request and response messages

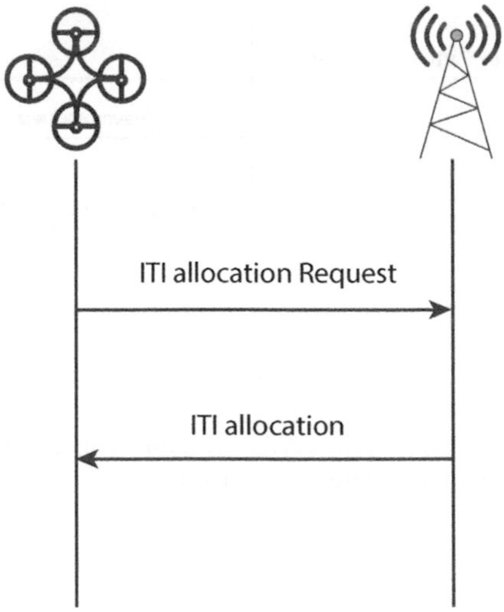

Figure 3.3 ITI allocation in UAV communication.

exchanged between a UAV and a controlling base station. The controlling base station generates and allocates a unique ITI to the UAV in response message. Note that simultaneous ITI allocation requests from multiple UAV (if any) can be resolved easily resolved as per the rules proposed in [43]. Here, the UAV can be either UAV-UE or UAV-relay or UAV-mini base station. In every case, the basic procedure would remain the same. Figure 3.2(b) can also be extended in context of ITI for UAV communications.

3.8 Decision Support System (DSS) in ITI Allocation

In order to employ ITI in UAV communications, there is a requirement of a decision support system (DSS). The DSS must enable allocation of interleavers for UAVs based on request, and it must be accountable for at least one interleaver allocation mechanism, such as ITI. The DSS works in coordination with the base station and can be an integrated part of the base station. Alternatively, it can be a separate system or an entity of the base station. Interleaver generation algorithm can be run on it. The DSS possesses some intelligence logic to allocate a better interleaver to a UAV so that interference remains at a minimum or zero level and interleaver allocation process always remains optimized. In more advanced cases, it may also include a machine learning algorithm in its decision logic for process optimization.

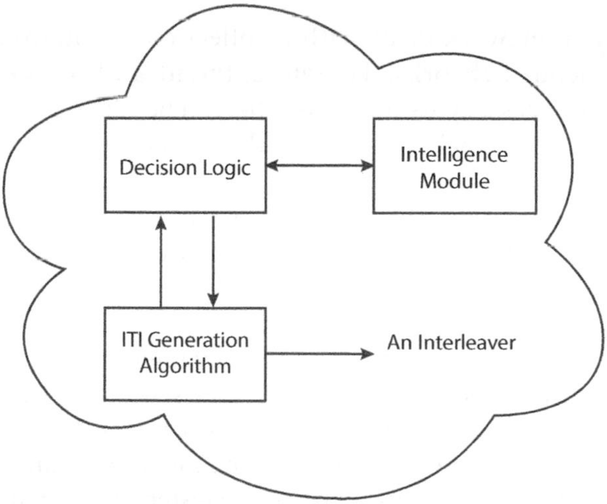

Figure 3.4 Basic DSS model for ITI allocation.

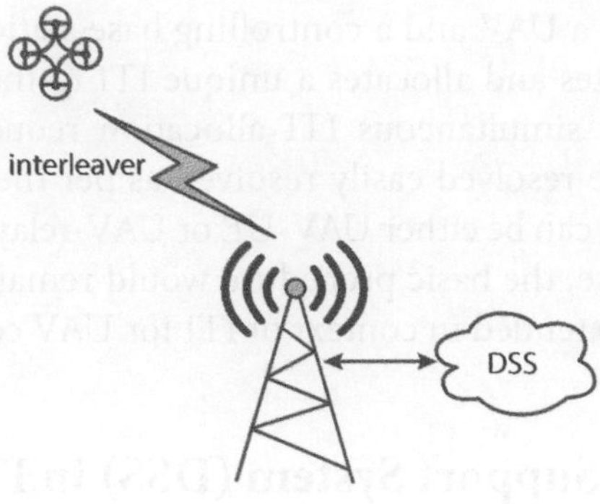

Figure 3.5 Coordination between base station and DSS.

Figure 3.4 illustrates elements of basic DSS model for ITI allocation in UAV communications. The DSS model has a decision logic that can communicate with ITI (algorithm) generator to make an informed decision for ITI allocation. Optionally, an intelligence module may be helpful in adding intelligence to prior to the decision. Machine learning algorithm can be a part of this module. Figure 3.5 represents the coordination between a base station and a DSS, where both the entities can be either two separate entities or single joint entity.

3.9 ITI-Based Clustered Interleaving and DSS for Smart City Framework

A smart city framework deals with a collection of multiple networks, i.e. heterogeneous networks. Therefore, the idea of "cluster interleaving" becomes necessary to be coined here. The cluster interleaving is an interleaving approach in which either same interleaver or different interleavers allocated to users located within a particular cluster (region). One type of interleaving pattern can serve users of a particular type of application service and other interleaving pattern can serve users of other particular service, and so on. These users may possibly exist in same cluster.

Figure 3.6 shows that cluster-1 is served by interleaver set-I (a1, a2, a3, a4, …) and cluster-2 is served by interleaver set-II (b1, b2, b3, b4, …), where (for e.g.) interleavers a1 and b1 can be same or different. Similarly, the other interleavers, for e.g., a1, a2 or b1, b2 can also be same or different.

By provisioning a DSS in ITI-based cluster interleaving approach for UAV communication within smart city framework, an efficient ITI

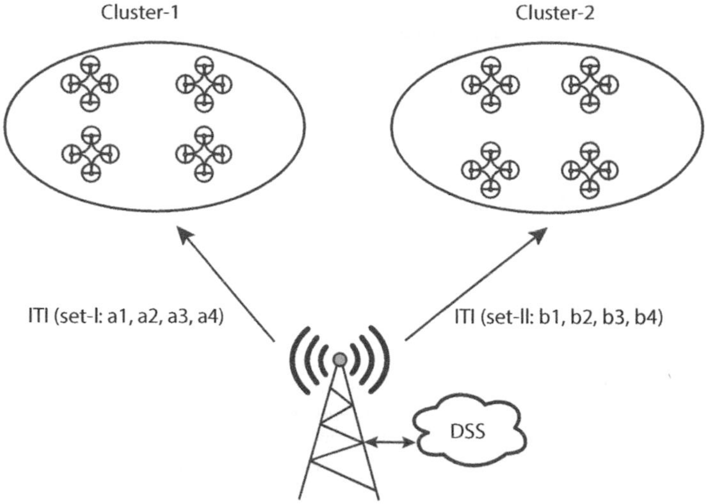

Figure 3.6 ITI-based cluster interleaving with DSS for smart city framework.

allocation is possible for the users of one or more clusters. Note that "ITI-based cluster interleaving" may also be termed as "FLRITI-based cluster interleaving" as ITI and FLRITI are synonyms of each other. In existing literature, it is quite evident that both the terms are equally popular. A smart city framework often needs an optimization to accommodate huge number of users of heterogeneous network. Smart interference mitigation would be possible only if an efficient DSS works at the network backend. Therefore, a DSS is essentially required for advanced applications to be served through ITI (or FLRITI) scheme.

Special Case: There could be a scenario when number of UAVs in a cluster is more than the number of interleavers in a particular set. For handling such scenario, the SS plays a key role. When the DSS predicts using its intelligence that such situation may arise or has already been occurred, the DSS allocates some of the reserved (free) interleavers to the cluster which requires it. These interleavers are allocated temporarily for a definite period only and then, these are released again and reserved again for future uses. Hence, reserved interleavers are those that are not dedicatedly allocated to any cluster. They are allocated by the DSS on demand and/or network requirement basis. Figure 3.7 depicts the role of DSS in allocating temporary interleaver, i.e., ITI. This figure is self-explanatory; hence, it does not need any further description.

This provision facilitates to serve special 5G use cases such as ultra-reliable low latency communication (URLLC), emergency communications, etc.

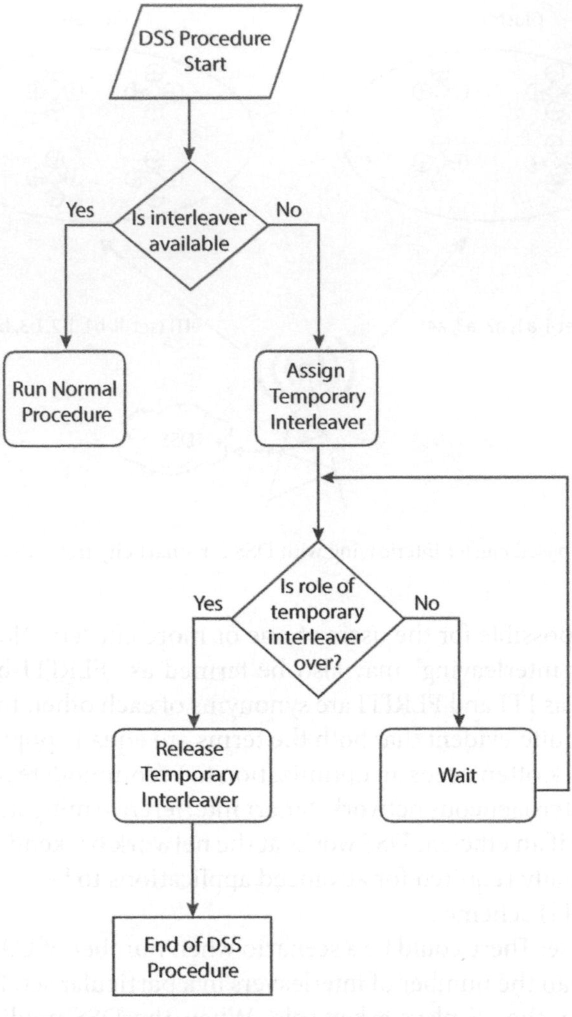

Figure 3.7 Role of DSS in allocating temporary interleaver/ITI.

An alternative solution to this scenario could be the borrowing of interleavers from the other cluster. However, this will not be possible in every circumstance. Therefore, the role of DSS is really important.

Enabling DSS for smart cities: - The DSS with intelligent ITI or any other interleaver selection capability can be used within smart city framework also. As in a smart city, everything is a part of a huge heterogeneous network. This huge heterogeneous network is formed with the collection of small sub-networks [45–51]. Every sub-network can be visualized as an independent network facilitating one specific application. This can be accessed by any user. Therefore, the DSS can categorize certain cluster of ITIs or interleavers for one particular application and other clusters for other applications.

3.10 Conclusion

All UAVs depend on their battery power for their operation and therefore, their limited power is a constraint in their operations. Another constraint for UAVs is interference from other nearby UAVs. Interleavers such as ITIs or random interleavers can be a great help in this direction. The theoretical concepts introduced in this chapter are meant for power saving and mitigating interference in UAV communications. The novel introduced concept of clustered interleaving is applicable in UAV communications and other general applications including smart cities framework. The decision support system, i.e., DSS roles have been also been elaborated in the chapter for both interleaver selection and making smart contextual decisions, wherever required. Interleaver centric approach for interference mitigation in UAV communication opens a new pathway for future researchers, and it requires their further thoughtful consideration to bring this into implementation phase for commercial usage.

References

1. Gruber, M., Role of altitude when exploring optimal placement of UAV access points, in: *Proceeding of IEEE Wireless Communications and Networking Conference*, Doha, pp. 1–5, 2016, https://doi.org/10.1109/WCNC.2016.7565073.
2. Yang, L., Juntong, Q., Xiao, J., Xia, Y., A literature review of UAV 3D path planning, in: *Proceeding of the 11th World Congress on Intelligent Control and Automation*, Shenyang, pp. 2376–2381, 2014, https://doi.org/10.1109/WCICA.2014.7053093.
3. Chen, J., Zhou, Y., Lv, Q., Deveerasetty, K.K., Ugochi Dike, H., A review of autonomous obstacle avoidance technology for multi-rotor UAVs, in: *Proceeding of IEEE International Conference on Information and Automation (ICIA)*, Wuyishan, China, pp. 244–249, 2018, https://doi.org/10.1109/ICInfA.2018.8812473.
4. Kim, J., Kim, S., Ju, C., Son, H., II, Unmanned aerial vehicles in agriculture: A review of perspective of platform, control, and applications. *IEEE Access*, 7, 105100–105115, 2019. https://doi.org/10.1109/ACCESS.2019.2932119.

5. Xu, C., Liao, X., Tan, J., Ye, H., Lu, H., Recent research progress of unmanned aerial vehicle regulation policies and technologies in urban low altitude. *IEEE Access*, 8, 74175–74194, 2020. https://doi.org/10.1109/ACCESS.2020.2987622.
6. Shakhatreh, H. *et al.*, Unmanned aerial vehicles (UAVs): A survey on civil applications and key research challenges. *IEEE Access*, 7, 48572–48634, 2019.
7. Davies, L., Bolam, R.C., Vagapov, Y., Anuchin, A., Review of unmanned aircraft system technologies to enable beyond visual line of sight (BVLOS) operations. *2018 X International Conference on Electrical Power Drive Systems (ICEPDS)*, Novocherkassk, pp. 1–6, 2018.
8. Mahjri, I., Dhraief, A., Belghith, A., A review on collision avoidance systems for unmanned aerial vehicles, in: *Nets4Cars/Nets4Trains/Nets4Aircraft 2015: Communication Technologies for Vehicles, Lecture Notes in Computer Science*, vol. 9066, Springer, Cham, 2015. https://doi.org/10.1007/978-3-319-17765-6_18.
9. Virgil, C., The use of decision support systems (DSS) in smart city planning and management. *Rom. Econ. Bus. Rev. Romanian-American Univ.*, 8, 2, 238–251, December 2014.
10. Ahmad, F., Decision support systems in a smart city: A review. *Proceedings of International Conference on Industry 4.0: A Global Revolution in Business, Technology & Productivity*, Segi University, Malaysia, October 14-16 2019.
11. Liu, Y. *et al.*, Unmanned aerial vehicle for internet of everything: Opportunities and challenges. *Comput. Commun.*, 155, 66–83, April 2020. https://doi.org/10.1016/j.comcom.2020.03.017.
12. Uragun, B., Energy efficiency for unmanned aerial vehicles. *10th International Conference on Machine Learning and Applications and Workshops*, Honolulu, HI, pp. 316–320, 2011, https://doi.org/10.1109/ICMLA.2011.159.
13. Amoiralis, E., II, Tsili, M.A., Spathopoulos, V., Hatziefremidis, A., Energy efficiency optimization in UAVs: A review. *Mater. Sci. Forum*, 792, 281–286, 2011.
14. Boukoberine, M.N., Zhou, Z., Benbouzid, M., Critical review on unmanned aerial vehicles power supply and energy management: Solutions, strategies, and prospects. *Appl. Energy*, 255, 113823, 2019. https://doi.org/10.1016/j.apenergy.2019.113823.
15. Yadav, M., Shokeen, V., Singhal, P.K., Flip left-right approach based novel inverse tree interleavers for IDMA scheme. *AEU–Int. J. Electron. Commun.*, 81, 182–191, 2017. https://doi.org/10.1016/j.aeue.2017.07.025.
16. Yadav, M., Shokeen, V., Singhal, P.K., Flip left-to-right approach based inverse tree interleavers for unconventional integrated OFDM-IDMA and SCFDMA-IDMA systems. *Wireless Pers. Commun.*, 105, 1009–1026, 2019. https://doi.org/10.1007/s11277-019-06133-3.
17. Bork, D., Buchmann, R., Igor, H., Karagiannis, D., Tantouris, N., Walch, M., Using conceptual modeling to support innovation challenges in smart cities.

2016 IEEE 18th International Conference on High Performance Computing and Communications; IEEE 14th International Conference on Smart City; IEEE 2nd International Conference on Data Science and Systems (HPCC/SmartCity/DSS), pp. 1317–1324, 2016.

18. Bartolozzi, M., Bellini, P., Nesi, P., Pantaleo, G., Santi, L., A smart decision support system for smart city. *Proceedings of IEEE International Conference on Smart City/SocialCom/SustainCom (SmartCity)*, Chengdu, China, pp. 117–122, December 19-21 2015, https://doi.org/10.1109/SmartCity.2015.57.
19. Yadav, M., Shokeen, V., Singhal, P.K., *Design and Analysis of Performance Enhanced Novel Interleavers for IDMA*, PhD Thesis, Amity University, Noida, India, 2019.
20. Hamaguchi, K., Ma, Y., Takada, M., Nishijima, T., Shimura, T., Telecommunication systems in smart cities. *Hitachi Rev.*, 61, 3, 152–158, 2012. http://www.hitachi.com/rev/pdf/2012/r2012_03_107.pdf.
21. Jawhar, I., Mohamed, N., Al-Jaroodi, J., Networking architectures and protocols for smart city systems. *J. Internet Serv. Appl.*, 9, 2, 1–16, 2018. https://doi.org/10.1186/s13174-018-0097-0.
22. Maitakov, F.G., Merkulov, A.A., Petrenko, E.V., Yafasov, A.Y., Development of decision support systems for smart cities, in: *EGOSE 2018: Electronic Governance and Open Society: Challenges in Eurasia, Communications in Computer and Information Science*, vol. 947, Springer, Cham, 2019, https://doi.org/10.1007/978- 3-030-13283-5_5.
23. Ling, B., Dong, C., Dai, J., Lin, J., Multiple decision aided successive interference cancellation receiver for NOMA systems. *IEEE Wireless Commun. Lett.*, 6, 4, 498–501, 2017. https://doi.org/10.1109/LWC.2017.2708117.
24. Fouda, A., Ahmed, S., II, Ismail, G., Ghosh, M., Interference management in UAV-assisted integrated access and backhaul cellular networks. *IEEE Access*, 7, 104553–104566, 2019. https://arxiv.org/pdf/1907.02585.pdf.
25. Yujie, T., Yang, P., Wu, W., Mark, W., Shen, X., Interference mitigation via cross-tier cooperation in heterogeneous cloud radio access networks. *IEEE Trans. Cogn. Commun. Netw.*, 6, 1, 201–213, 2020. https://doi.org/10.1109/TCCN.2019.2957457.
26. Yatong, W., Gang, F., Yao, S., Qin, S., Ying-Chang, L., Decentralized learning based indoor interference mitigation for 5G-and-beyond systems. *IEEE Trans. Veh. Technol.*, 69, 10, 2124–12135, 2020. https://doi.org/10.1109/TVT.2020.3012311.
27. Ping, L., Liu, L., Wu, K., Leung, W.K., Interleave division multiple-access. *IEEE Trans. Wirel. Commun.*, 5, 4, 938–947, 2006. https://doi.org/10.1109/TWC.2006.1618943.
28. Wu, K.Y., Ping, L., Liu, L., Leung, W.K., Interleave division multiple access (IDMA) communication systems, in: *Proceedings of 3rd International Symposium on Turbo Codes and Related Topics*, Brest, France, pp. 173–180, 2003.

29. Vázquez-Araújo, F.J., González-López, M., Castedo, L., Garcia-Frias, J., Interleave-division multiple access (IDMA) using low-rate layered LDGM codes. *Wirel. Commun. Mob. Comput.*, 12, 14, 1276–1283, 2012. https://doi.org/10.1002/wcm.1055.
30. Yadav, M., Gautam, P.R., Shokeen, V., Singhal, P.K., Modern fisher-yates shuffling based random interleaver design for SCFDMA-IDMA system. *Wireless Pers. Commun.*, 97, 1, 63–73, 2017. https://doi.org/10.1007/s11277-017-4492-9.
31. Shokeen, V., Yadav, M., Singhal, P.K., Comparative analysis of FLR approach based inverse tree and modern fisher-yates algorithm based random interleavers for IDMA systems. *IEEE 8th International Conference on Cloud Computing, Data Science & Engineering (Confluence)*, Noida, pp. 447–452, 2018. http://doi.org/10.1109/CONFLUENCE.2018.8442676.
32. Yadav, M., *An Investigation on Interleaving and Interleavers for Multiple Access Systems. A Tutorial*, Grin Verlag, Munich, 2020. https://www.grin.com/document/935040.
33. Yadav, M., *State-of-the-Art. Theory of Interleavers*, Grin Verlag, Munich, 2020. https://www.grin.com/document/933425.
34. Yadav, M. and Singhal, P.K., Chapter-5 Interleavers-centric conflict management solution for 5G vehicular and cellular-IoT communications, in: *Cloud and IoT Based Vehicular Ad Hoc Networks*, G. Singh, V. Jain, J.M. Chatterjee, L. Gaur (Eds.), pp. 83–104, Scrivener Publishing LLC. [Accepted: In Press], 2021. https://doi.org/10.1002/9781119761846.ch5
35. Agarwal, P. and Shukla, M., MITA interleaver for integrated and iterative IDMA systems over powerline channel. *Wireless Pers. Commun.*, 122, 1559–1575, 2022. https://doi.org/10.1007/s11277-021-08961-8
36. Hassija, V., Saxena, V., Chamola, V., A blockchain-based framework for drone-mounted base stations in tactile internet environment. *IEEE INFOCOM: Computer Communications Workshops (INFOCOM WKSHPS)*, pp. 261–266, 2020. https://doi.org/10.1109/INFOCOMWKSHPS50562.2020.9162991.
37. Yadav, M., Shokeen, V., Singhal, P.K., Uncoded integrated interleave division multiple access systems in presence of power interleavers. *Radioelectron. Commun. Syst.*, 60, 11, 503–511, 2017. https://doi.org/10.3103/S073527271711005X.
38. Yadav, M., Shokeen, V., Singhal, P.K., BER versus BSNR analysis of conventional IDMA and OFDM-IDMA based systems with tree interleaving, in: *IEEE 2nd International Conference on Advances in Computing, Communication and Automation (ICACCA)-Fall'16*, pp. 1–6, 2016. http://doi.org/10.1109/ICACCAF.2016.7748973.
39. Yadav, M. and Banerjee, P., Bit error rate analysis of various interleavers for IDMA scheme, in: *Proceedings of the IEEE 3rd International Conference on Signal Processing and Integrated Networks (SPIN)'16*, Noida, India, pp. 89–94, 2016. https://doi.org/10.1109/SPIN.2016.7566668.

40. Shokeen, V., Yadav, M., Singhal, P.K., Simulation performance of conventional IDMA system with DPSK modulation and modern fisher–yates interleaving schemes, in: *Smart Computational Strategies: Theoretical and Practical Aspects*, A.K. Luhach, K.B.G. Hawari, I.C. Mihai, P.A. Hsiung, R.B. Mishra, (Eds.), Springer, Singapore, 2019. https://doi.org/10.1007/978-981-13-6295-8_14.
41. Agrawal, P. and Shukla, M., Effect of various interleavers on uncoded and coded OFDM-IDMA over PLC. *2020 IEEE In Proceedings of the Fifth International Conference on Communication and Electronics Systems (ICCES 2020)*, Coimbatore, India, pp. 275–279. https://doi.org/10.1109/ICCES48766.2020.9137902.
42. Shukla, M., Srivastava, V.K., Tiwari, S., Analysis and design of optimum interleaver for iterative receivers in IDMA scheme. *Wirel. Commun. Mob. Comput.*, 9, 10, 1312–1317, 2009. https://doi.org/10.1002/wcm.710.
43. Yadav, M. and Singhal, P.K., Simultaneous interleaver assignment requests handling in inverse tree interleaver deployment scenario for multi-user 5G communications. *Wireless Pers. Commun.*, 118, 1129–1147, 2021. https://doi.org/10.1007/s11277-020-08062-y.
44. Yadav, M. and Singhal, P.K., Interleaver assignment solution for multiradio access technology supported 5G networks. *Radioelectron. Commun. Syst.*, 64, 99–105, 2021. https://link.springer.com/article/10.3103/S0735272721020059.
45. Linger, R.C. and Hevner, A.R., Flow semantics for intellectual control in IoT systems. *J. Decis. Syst.*, 27, 2, 63–77, September 2018. https://doi.org/10.1080/12460125.2018.1529973.
46. Impedovo, D. and Pirlo, G., Artificial intelligence applications to smart city and smart enterprise. *Appl. Sci.*, 10, 8, 2944, April 2020. https://doi.org/10.3390/app10082944.
47. Ribeiro, S.S., Rezende, D.A., Jingtao, Y., Toward a model of the municipal evidence-based decision process in the strategic digital city context. *Inf. Polity*, 24, 3, 305–324, August 2019. https://doi.org/10.3233/IP-190129.
48. Nesi, P., Badii, C., Bellini, P., Cenni, P., Martelli, G., Paolucci, M., Km4City smart city API: An integrated support for mobility services. *Proceedings of IEEE International Conference on Smart Computing (SMARTCOMP)*, St. Louis, MO, USA, pp. 1–8, May 18-20 2016. https://doi.org/10.1109/SMARTCOMP.2016.7501702.
49. Haidine, A., Hassani, S.E., Aqqal, A., El Hannani, A., *The Role of Communication Technologies in Building Future Smart Cities*, I.N. Da Silva and R.A. Flauzino (Eds.), IntechOpen, 2016. https://doi.org/10.5772/64732.
50. Yadav, M., Shokeen, V., Singhal, P.K., Testing of Durstenfeld's algorithm based optimal random interleavers in OFDM-IDMA systems. *3rd IEEE International Conference on Advances in Computing, Communication & Automation (ICACCA) (Fall)*, Dehradun, pp. 1–4, 2017. http://doi.org/10.1109/ICACCAF.2017.8344675.

51. Yadav, M., Shokeen, V., Singhal, P.K., Bit error rate analysis of DRI centric IDMA and IIDMA systems with underwater channel assumption. *2018 IEEE 8th International Conference on Cloud Computing, Data Science & Engineering (Confluence)*, Noida, pp. 411–416, 2018. http://doi.org/10.1109/CONFLUENCE.2018.8443041.

4

Introduction to DSS System for Smart Cities

Kuldeep Singh Kaswan[1*], Ravinder Gautam[2] and Jagjit Singh Dhatterwal[3]

[1]*School of Computing Science and Engineering, Galgotias University, Greater Noida, India*
[2]*Maharshi Dayanand University, Rohtak, Haryana, India*
[3]*Department of Artificial Intelligence & Data Science, Koneru Lakshmaiah Education Foundation, Vaddeswaram, AP, India*

Abstract

The smart city infrastructures meet new difficulties to enhance the consistency and the sustainability of facilities for the resident and have additional characters to help the ecosystem of the urban sector to adaptively configure according to collected knowledge may, therefore, not be ads. In this context, Decision Support Systems (DSS) recently became increasingly relevant. It consists of a smart decision support system for smart city based on the development of the model of the analytical hierarchical process integrated into the logic representation system. A basic example of using the Convolutionary Neural Network (CNN), which can be used for automatic fire detection to provide a suitable reaction, has also been developed. In order to find flaws, minimize risks, and establish tighter security measures than are already in place to deter cyberattacks, this scheme would propose linking to Open-Source Intelligence (OSINT). The DSS components provide strategic, logistical and organizational decision makers with the resources they need to schedule and complete the collection of projects necessary to create a smart town and/or smart house. DSS is positioned to be a unique benefit as a constantly evolving tool to prepare and complete the exponential growth, both "IT, Cloud Computing, the web-based stuff, everything-as-a-service (XaaS) and the development of new mathematical models, artificial intelligence and data storage." This chapter presents decisions, decision modeling and decision management principles, support mechanisms for decision making, and collaborative systems, as well as how they can assist in a planning initiative in the smart city.

**Corresponding author*: kaswankuldeep@gmail.com

Loveleen Gaur, Vernika Agarwal and Prasenjit Chatterjee (eds.) Decision Support Systems for Smart City Applications, (53–76) © 2023 Scrivener Publishing LLC

***Keywords*:** Decision support systems, information society, ICT platform, Internet of Thing, everything-as-a service, smart city, web-based DSS

4.1 Introduction

The change from a conventional to a smart city takes place by facilitated efforts to create a more smoothly, efficiently and safely controlled city [1]. This has been achieved by convergent policies coping with Smart Transit Networks, Ecosystems, Information and Communications Technology (ICT) Assistance, Smart Public Safety, Health Systems, Environmental Sustainability and Positive Energy. In the last ten years, the Smart City idea has been created. Several companies, Substantial investment in sustainability urban projects that use ICT and IoT to analysis, prepare and build solutions are made possible by education universities, cities and governments (Information and Communication Technologies and easy access to Internet-of-Things) [2]. IBM Intelligent Planet, Oracle I-Government, Austin, Barcelona, Helsinki, Amsterdam, smart city and Dubai, Dubai, and smart cities are some of the highlights of innovative Smart City projects. The quickly anticipated rise of cloud infrastructure and the XaaS industry would also provide high demand and encouragement for conventional town transformation into a smart community.

The creation of a modern Information Society has been made possible by information technology, particularly the internet, mobile telecommunications and simple Internet of Things (IoT). In several areas such as public health, traffic safety, e-commerce, and energy, information and communication technology (ICTs) are now being used. In the process of converting the traditional town into a smart town, the knowledge society plays an important role. In order to enhance quality of life for the European resident, the European Union (EU) has established several initiatives in different fields for the use of ICT facilities. In the following area the particular large-scale ITC projects were invented [3]:

- Road security: e-call, i2010 Smart Car Project, Emergency in Cars, e-security;
- Economic Trading: "Electronic Trade Directive;"
- Payment processes: Group network of electronic signatures, automated transfers, the relationship with the card's supplier;
- New Tech: A Search policy on Europe's future and new technologies;

- Oil industry: ICTs for promoting the transition to a low-carbon and energy efficient economy;
- Public health: programs and facilities in telemedicine [4].

4.2 Smart City System Architecture

Like intelligent transportation, smart health, intelligent clothes, intelligent grid, wearable electronics and intelligent services, to name a few [5]. Although these applications vary considerably from one typical operating architecture to another, nearly all can be analyzed in five different planes: (i) application, (ii) sensing, (iii) connectivity, (iv) data and, (v) protection planes. The remainder of this section deals with various levels of intelligent town design. Section 4.8 explores a selected number of intelligent city systems from a sensing, connectivity and applications perspective [6].

- Application Plane: - The software is the gateway between an autonomous city and its users (mostly its residents). Applications are designed to minimize city expenditures by managed resource theft, work automation, and omnivore and continued surveillance to increase protection and security of the city [7].

4.2.1 Sensing Plane

A standard sensor level integrates a broad number of sensing instruments and actuators in order to calculate and communicate with physical signals (e.g., environmental irradiation) (e.g., city lights). The deployment of sensing aircraft in smart towns is comparable to conventional Wireless Sensor Networks (WSNs) with similar constraints. Most importantly, low supply of electricity remains a big barrier. This constraint is not therefore induced, unlike WSNs, by the lack of the power delivery system, but also by the large-scale of the aircraft [8].

4.2.2 Communication Plane

In the contact planes, the data obtained from the sensor plane is preprocessed and applied and transmitted to the other layers so that field instruments can be easily connected to the cloud. Preferably, this connection can enable low latency, high performance, scalable, and stable communication. However, a variety of obstacles preclude such a conceptual contact service from being practicable. This forces programmers to make choices however according their frameworks' specifications [9].

The contact capacity derived from the WSNs (IoT's ancestor) is primarily inhibited by the non - availability. In the contact plane (as opposed to the data plane), the consequences of this shortage are much greater, since the communications circuit absorbs more energy than sensing orders of magnitude. Although approaches to capture energy from the ambient will mitigate this negative impact in certain situations, it is the most feasible approach to sacrifice the battery life of sensors by supplying data speed, latency and transmitting (which host the front-end circuitry for communication) [10].

4.2.3 Data Plane

A data plane is the point of convergence of the collected data which transforms the plurality of apparently incoherent information into logical ones. Often, these approaches entail sophisticated algorithms, which requires excellent computerized hosts [11]. In comparison, the large volume of IoT and the short time limit for certain intelligent city systems e.g., smart grid, transport and health services) underline the value of computing resources. No single node is capable of meeting this obligation in the data and connectivity planes. However, individual capabilities can be merged as one abstract computing infrastructure (e.g., a cluster) in which each node performs part of an algorithms. By optimizing the services available for the delivery (or decentralization), recurrent and non-recurring costs can be considerably minimized [12]. The feasibility of the approach is specifically dependent on the algorithm and its distributed execution capacity. For the alternative, the data plane may be centrally designated in geographically isolated, clouded connections to provide dynamic output and interaction functionality. Although costlier, cloud-based servers are normally more robust and able to perform traditional off-shelf algorithms [13].

4.2.4 Security Plane

Notwithstanding several developments in IoT, the field remains in its infancy; a number of immatures are being implemented simply to assess the viability and adoption by consumers of different concepts. Privacy concerns are frequently ignored in such an environment. Recent times, although the significance of IoT privacy and security in the field has recently been increasingly recognized through complex cyber-attacks in the IoT arena. While most of these attacks are targeted at ransom on a smaller scale, there are continued concern that big attacks will hit vital infrastructure that can affect the economy and threaten the lives of several smart urban residents [14].

4.3 Types of Network Sensors

Of course, without the use of sensors, a smart city is unlikely. A sensor for measuring the physical property of any entity or condition is used in the smart town. Biosensors, electrical sensors, chemical sensors, and intelligent grid sensors are the major sensors used. In the subsequent paragraphs, the forms and functions of these sensors are discussed [15].

4.3.1 Electronic Sensors

Smart devices primarily utilize the electroscope, electromagnetic disturbances and voltage sensors as a way of detecting multiple types of electricity. Smart devices involve environmental monitoring sensors, parking sensors, speed-meters, etc. The wireless characteristic of the sensors indicates that energy is strong, versatile, and highly complex [16]. The intelligent sensor is capable of understanding and analyzing videos from various sensors, including street surveillance cameras and cell phones. Its high precision in information collection and conversion from acts, such as gesture movements, contributes to its implementation of intelligent services in computer interactions [17].

4.3.2 Chemical Sensors

Material sensors such as atmospheric carbon sensors, oxygen, mechanical noses and catalyzed ball sensors are actually used in the chemical sensor. When interacting with chemical substances, all the issues associated with chemical substances are very satisfactory to be found. In other words, chemical change and the chemical composition of the structure or atmosphere are useful for the identification [18].

4.3.3 Biosensors

The biosensor's primary purpose is to identify the biomedical analyses. Among several other biosensor instruments, existing biosensors provide sensors for ionizing and sub-atomic sensors such as neutrons and MEMS. The biosensors are limited so they can only function in the field of biomedicine [19].

4.3.4 Smart Grid Sensors

In terms of function, sensors, innovative modules, connectivity, decision support systems and certain other highly enhanced component parts, smart grid innovations are divided into five main segments [20]. Smart grids use different sensors to efficiently generate, transmit and distribute energy from the power generation network to the application developers.

4.4 Role of Sensors in Smart Cities

Today sensors are playing the main role in various domains of a smart city such as security, safety, traffic control, parking, transportation, environment, etc. All these are discussed below.

4.4.1 Safety and Security Management

Despite the substantial advancement in emergency medical care, cities frequently risk their lives due to gaps in reporting and reaction time. It is also vital to reflect not only on crime reduction and emergency preparedness but also on solutions for mitigating. Sometimes only after incidents were the police or the firemen talked about injuries or offences. The installation of gunshots, crash and noise alarms in the city would reduce this delay. Able to connect smoke alarms to a central artificial intelligence device will also support police, ambulances and firefighters respond rapidly [21].

4.4.2 Service Delivery and Optimization

Illumination and plumbing systems for large communities were confronted by local authorities. Manual and primitive control mechanisms are used by some municipalities, which are vulnerable to mistakes and can obstruct the optimization process of management. The probability of simplifying service delivery is that an automated framework in a smart city that uses a higher-number sensor node than the city's population [22]. As data are consolidated from these sensors, they will promote the everyday activity and maintenance of intelligent cities.

4.4.3 Traffic Control and Parking

Because of road jams, cities lose millions of dollars. Time and fuel were wasted on highways as a result of temporary immobilization. Moreover, the quality of living in cities declines. In addition, it is clear that it is difficult to locate parking spaces that is causing the congestion. In addition, in an era of electric vehicles transport control is crucial; the use of smart metering tools therefore solves the problem effectively [23].

4.4.4 Smart Building

Building administration is left to independent property owners in conventional schemes. It is also possible to make processes effectively unpredictable, as is the expense of hiring and retaining property managers high. For example, documenting and responding to the maintenance needs of tenants can take many days. The conventional systems often require a guard to control the entrance and exit of a house, thereby improving protection. The problems raised by conventional structures can be eliminated with intelligent buildings. Smart cities can take advantage of the benefits of robotics that is essential to intelligent design. Automation in intelligent buildings should provide a decentralized infrastructure which can be accessed.

4.4.5 Public Transport

The bulk of public transit services run at bus stations consistently. Daily stops can lead to loss of time, though, as there may be no switches. In comparison, time spent maneuvering by buses around areas beyond the specified route is an issue that often does not distinguish future travelers. This means that a system of linked terminals at all bus stations and bus dashboards will be built to eliminate these contradictions. This device will help drivers gather data to ensure performance. The implementation of an information system linked to remote controls in smart cities is needed to incorporate this solution. In monitoring the progress of buses, GPS may be invaluable.

4.4.6 Environment

Smart sensors can increase deterioration of the climate. However, it is possible to solve the climate change problems facing clever cities by developing and efficiencies in electricity production and use. In order to acquire accurate information, the main challenge is to handle broadly used sensors [24].

4.4.7 Ethical Implications

In the introduction of intelligent cities, safety and protection are the key ethical concerns. Failure to ensure safety leads, among other factors, to abuse and theft. The legal ramifications of intelligent networks and cities are demonstrated. Ethical issues play a vital role for all parties, including the government and politicians.

4.5 Implications of Smart Sensors

The following implications of the intelligent sensor functions are suggested on the basis of the findings of the exploratory performed above.

First, the reduction of incident detection and response times would increase prevention and protection. In addition, IoT can resolve safety and protection issues by linking smoke and noise sensors.
Second, the sensors can be used for consolidating waste data and repair sensors for improved service delivery.
Third, in directing traffic signals may be used; therefore, congestion can be removed and parking areas handled.
Fourth, in the construction of autonomous cities, smart networks can be beneficial. There are integrated buildings with facilities and resources operated.
Fifth, we are discussing the environmental and public transit consequences of intelligent cities.
Sixth, we plan to discuss the overall ethical concerns around intelligent sensors, including data security and the regulatory system. These guidelines continuously accumulate and direct the process of implementation.

Finally, we strive to achieve the task of applying solutions by showcasing ways to address challenges through the use of network sensors in intelligent cities.

4.6 Decision Modeling

Three fundamental factors: matter, strength, and knowledge are affected by any human action and mission. The flow of knowledge associated with various activities/tasks in the city has two major elements: one future (related to forecasts, preparation and project planning) and one longitudinal (related to track and handle tasks and activities).

- **Types of Decisions** There are three tiers of management: political, tactical and organizational levels in all organizations (simple, medium or large) and smart cities. The right decision making at each level is a required prerequisite for the performance of the enterprise (most definitely in useful time). Decisions may be rated on the basis of the essence of decisions themselves (programmed based not) or on a decision-making basis (strategic, tactical and operational level) at organizational management level:
    - Regular and repeated decisions are scheduled, and the company usually establishes unique ways of handling the decisions. A programmed decision may include how items are arranged on the store shelf.
    - Policy choices – which affect the identity and method of delivering long-term or global goals and interests;
    - Operational/Organizational decisions – are associated with the coordination and effectiveness of various facets and divisions of a company.
    - Organizational decisions – contribute to the everyday running of a group or entity.
- **Modeling of decisions and procedures for decision making:** The simulation of actions makes the logic possible of business evident recognizable, improving the comprehension of the interactions between business logic, procedures and knowledge by each management level. When it is used to complete a good decision-making process, the reality becomes information.

  Decision theory notes that the following components mark the decision-making process:
    - The criterion for judgment (the questions are evaluated from various points of view);
    - goal (or targets to be pursued);
    - An individual or group of individuals who would like to make the right decision to achieve the goals and objectives.

- The set of alternatives (includes all of the practicable steps to attain the goal);
- A collection of possible states (the Many factors that decide the presence of any effect on a given objective in each state);

- **Decision Tree and Influence Diagram:** Decision tree algorithm are an algorithm and frequently use them as a method of evaluating the approach that is likely to achieve a target in organizational science, in particular in decision processes. Decision trees are informally useful when a community has to make a decision to concentrate on the issue. A systematic way of the estimation of conditional probability is another application of decision tree. Decision tree algorithm can even be added, and the solution or rendered using a professional software or graphical program [23].
- **Executive Information System:** Taking decisions and modelling decisions involves vast volumes of Helpful time for data and information. The management of information manages information exchange from every company, businesses (small, medium, large) and smart cities. The information systems are an outstanding solution. The Executive Information System (ECIS), which merged knowledge within the enterprise with information resources in an empirical context, is a collection of strategic resources to serve the knowledge and decision-making requirements of the enterprise. Usually, EIS components can be defined as [24]:
  - Hardware (data entry machines, CPUs, data collection files and output devices);
  - Hardware; tools (text foundation, database, basic graphics, DSS);
  - User interface (planed reports, queries, menu oriented and input/output languages, command language); Telecommunications, which is essential in order to create a secure network (transmission of data from one location to another);
  - The applications device (EIS has been extended to many fields, in particular to development, marketing, distribution and financing).

4.7 Decision Support Systems (DSS)

An automated information service support framework for decision support allows users to make reasonable decisions on the basis of a list of customer criteria. The DSS is an information structure category serving enterprise and organization. An integrated program is a well-designed DSS framework designed for policy makers to compile valuable information from the raw data, records, personally identifiable information, or business strategies for problem solving and decision making.

4.7.1 Decision Support System Components

Data maintenance, concept creation, management of knowledge bases, dialog system generation, and implementation and a user interface can be included in the decision support systems. The sub-system Data Management (DM) gathers and section represents. The use of a master database and the structure for data base management requires data management.

Examples of DSS: As described in the overview, a project's planning and execution needs time, resources, staff, infrastructure and a lot of details, mostly significantly.

Present activities: Start-up tasks, Start-up tasks, Development tasks, finished tasks, sliding tasks;

- Cash flow, preparation, over-budgeting over-budget, value earned, weekly;
- Terms: who does what, who does what frequently and over-allocated resources are to be listed;
- Workload: operations utilization, resource utilization;
- Customs: Foundation plan, budget update, tasks done, cross-talk, benefit, development, capital overwhelmed, project outline, resources, resource use, begun tasks, sliding tasks, Success Tasks, beginning tasks, Top level Tasks, Unstartled Works, Weekly Cash Flow, Who Does Who.

4.7.2 DSS Merits and Demerits

In decision-making process, some of the benefits of DSS include:

Time saves - time savings, improved efficiency, shortened decision cycle time and timelier decision-making details [24].

- Improving performance—increased efficiency of decision formation and decision-making;
- Enhancing contact with people—can promote communication and decision-making collaboration;
- Comparative advantages for market intelligence, success improvement and web-based DSS systems;
- Encourage studying—learning new ideas and gaining a greater view of the corporate landscape and decision making;
- Improve corporate governance—Data-driven DSS also makes data used for performance analysis and ad hoc querying of company transactions.
- While DSS may provide unquestioning advantages and benefits for organizations, it really can also produce negative effects for those conditions by constructing and using DSS.

4.8 Chandigarh: Becoming a Smart City

Chandigarh is on a list of 98 towns protected by the Smart City Initiative of the Indian Government. In Chandigarh, the French government has listed up to two milliards Euros in investment in the next few years as part of the Indian government's smart city program. Chandigarh is one of four cities. Chandigarh is a city and a region for peace in the northern part of India, which acts as the capital of the states of Punjab and Haryana. The city as a UT is governed entirely by the union government and does not belong to either territory. The villages of Mohali and Punjab are known as Tricity Chandigarh, and the adjacent villages of Mohali, and of Panchkula, Haryana. Chandigarh is located in northwestern India, near the foothills of the Himalayan zone of Shivalik. It is roughly 114 km^2 in surface area. The Haryana and Punjab States share their borders. The exact map of Chandigarh is 30.74°N 76.79°E. It has an overall height of 321 m (1053 ft).

4.8.1 Chandigarh City's SWOT Analysis

- **Strengths:**
  Smart city is an advanced digital and smart city. A good urban metropolis and ICT are required components of the notion of intelligent city, but are not enough to be labeled an intelligent city. Smart city is a progressive idea, requiring urban knowledge and management information systems interaction and coordination with Information communication

technology of future generation like Internet of Things (IoT), cloud technology, etc. Smart city is the electronic and smart city's evolution. In the position of the idea of the modern metropolis, the intelligent city is to be the Information Road using remote sensing (RS), GIS, and GPS to consistently collect and analyze data and to build 3D functions for data monitoring and decision making. Smart cities use intelligent data to exchange knowledge and insights based on the IT infrastructure, such as embedded devices, machine intelligence, and smart decision-making systems to efficient overall metropolitan cities.

- **Opportunities:**
  Intellectual capital is the gradual portion of the power requirement of smart cities, thus human resources involvement should really be the beginning point instead of mindlessly taking into consideration that ICT and technology growth immediately alter and develop cities. A smart city may be built on the detection and measurement of residents' acceptance of the smart city idea and the identity behaviors of people knowledgeable of situations in future. Intelligent inhabitants in intelligent cities are not only marked by their qualification and experience but also by the intersection of personal connections that are not linked to external problems outside "personal and professional life." Cloud technologies from intelligent individuals, via consciousness, social participation and learning, retraining, enhancement of capacities, innovation, adaptability, openness, ethnic and cultural heterogeneity and cooperation amongst participants, may tackle current urban challenges.
- **Weaknesses:**
  System of government as one of the core ingredients in encouraging smart cities is committed to ensuring enrollment and facility collaboration between various responsive organizations and stakeholders, including political, commercial and scientific communities, nonprofit and voluntary agencies, with a view to improving underpaid and overworked in serving residents. Such a management's meaningful essence should be predicated on a citizen-centered and citizens-led approach that allows residents availability to and productive involvement in the decision making and community/welfare payments of ideas, premonitions, strategic

goals, expectations, and business strategies for smart cities through the use of the e-government concept. Citizen-centered government needs both to manage the outputs of people with diverse elements of the standard of living and the monetary products of individuals.

- **Threats:**
  The discrepancy of the legislation and requirements on fiscal policies in the worldwide, regional and national fields cannot assist to further the smart city efforts. There is also an absence in the monitoring and checking of capital appreciation of suitable, statistical techniques and measurements. Government spending is decelerating declining public spending from the recession. Advances in capital technological fields such as stelvio cities has been reduced by the economic state, the inability of credit and requirements on investment banks to decrease risk exposure by creating stronger deposits bases, which restrict the financing costs accessible.

4.8.2 Challenges to Become the Smart City for Chandigarh

- **Health Care:** Chandigarh is the birthplace of healthcare in the country, but there are still no basic facilities and labor requirements. The city only has 43, with 80 dispensaries each with 15,000 inhabitants for almost 13 lakh citizens.
- **Education:** In 2016, Colleges of Chandigarh crossed the figure set by the College community Index that needs one college against such a lakh student. The laboratories have to be converted in the seven government colleges in the city into virtual laboratories although well equipped. Hostel seats in colleges are absent, while 42 and 46 university students are having issues with connectivity. Panjab University has taken the lead on international channels, but it is unable to support it. There are yet to be built smart classroom and modern hostels
- **Solid Waste Management:** Crores was expended in the purchasing of lifting garbage trucks, trolley bins and rehris, but the method of collection and segregation of waste was not introduced at household level and Sehaj Safai Kendras. So, waste which can be recycled is disposed of in open spaces. Bio-medical waste control in the area is also unsuccessful.

- **Regulating Traffic:** Because of the number of cars, 9.75 lakh, Chandigarh is an important obstacle, while two lakh vehicles use urban roads every day from the outskirts.
- **Transport Services**: Chandigarh operates only a local bus service. In most lines, the bus service is shut down by 9 pm. A user-friendly information system for the bus route escapes the area. For some destinations two busses may be changed, but no road plans for direct routes remain.

4.9 A Topology of Smart City Functions

4.9.1 Smart Economy (Competitiveness)

- The city is pushed by innovation and supports a university which focuses on state-of-the-art technology, not just for science, manufacturing and the industry, but also traditional culture, construction, planning, development, and the like.
- A clever city value and encourages new creativity and solutions.
- A smart city has shed light on business leadership.
- A smart city has several economic options for its residents.
- A clever city recognizes that every business functions locally.
- A clever city is equipped for internationalization difficulties and possibilities.
- Intelligent city explores, encourages and encourages economic cooperation.
- A smart city believes, acts immediately and challenges internationally.
- A smart city invests strategically on its critical infrastructure.
- A smart city creates and promotes national and engaging brands.
- A smart town emphasizes that economic progress be equitable and sustained (growth).
- A smart city is a place for visitors (tourism).
- A smart city has identified important criteria that are nationally competent.
- A smart city makes use of its resources and finds answers to issues.
- A smart city is tremendously productive.
- A smart city has a high labor market flexibility.

- A smart city embraces its wealthy information systems.
- In a smart city, people strive for sustainable management of natural resources and recognize that its economy would not operate indefinitely without it.

4.9.2 Smart People (Social and Human Capital)

In what they do effectively, smart individuals thrive.

- The high human development index (48) is available to intelligent individuals.
- The intelligent city incorporates its educational institutions into the whole urban life.
- It draws high human resources, such as employees of knowledge.
- A clever city has a high Graduate Inscription Rate and a high degree and skill level.
- The people choose to employ e-learning methods and life-time learning.
- In an intelligent city, people are extremely flexible and robust to environmental changes.
- Intelligent city residents are creative and are able to find innovative solutions to hard problems.
- Intelligent individuals are cosmopolitan, fully accessible individuals, and have a multicultural outlook.
- Smart employees are effectively engaged, working efficiently and easily, maintaining and managing their cities and rendering them increasingly viable.

4.9.3 Smart Governance (Participation)

- In its administration, a smart city emphasizes responsibilities, reactivity and transparency (ART).
- Big data, geographical decision support and associated geo-spatial technologies are used in a smart city in urban areas regional government.
- An intelligent city is continually innovating e-government to help all its inhabitants.
- A smart town continually enhances its capacity for efficient and efficient public services.

- Collaborative policy making, planned, financing, implementation and control processes in a smart city.
- A clever city has a defined plan and objectives known to everyone for sustainable urban growth.
- The smart city uses urbanization innovative development with the concentration on the combination of urban growth in economic, social and environmental aspects.
- The intelligent city is efficient, efficient and manageable.
- E-Democracy is a clever city to produce better results for everyone.
- A smart city includes a triple scale model, which includes a change in the responsibilities of government, Academy and business/industry.

4.9.4 Smart Mobility (Transport and ICT)

The following elements are included in Key Technology, the third main component of a Smart City system.

- A clever city focuses on human transportation and not only car mobility
- Walking and bicycling will be promoted by a smart city.
- A clever city has lively roads (at no additional cost).
- An intelligent city controls traffic and traffic problems properly.
- An intelligent city has fun (bike) pathways.
- A clever city is equally well-off.
- A smarts city has a mass transport system for rapid transportation, such as the underground, light subway, monorail or maximum efficiency "skytrain."
- An intelligent city will have an interconnected high mobility system connecting residential buildings, places of employment, entertainment areas and transportation notes.
- An intelligent town is going to live in dense population so that high-speed transport is accessible universally.

A smart city offers uninterrupted movement for persons with diverse capabilities (commonly misnamed, impaired).

4.9.5 Smart Environment (Natural Resources)

- A smart town lives with environment and safeguards its environment.
- A smart town is beautiful and strongly anchored in the natural environment.
- Smart cities appreciate their cultural resources, their personalized recommendations, their uniqueness and their surroundings.
- A smart city maintains and maintains the urban area's natural system.
- In a smart city the biodiversity in the city is taken up and supported.
- A smart city handles its natural resources more efficiently.
- A smart city offers individuals of all age's entertainment opportunity.
- Green town is a clever town.
- A clean town is an intelligent town.
- Smart cities have sufficient public green areas that are approachable to them.
- An open-plan living area in a clever city. In contrast to the interior living room of homes, outdoor living spaces are personal, lively and dynamic urban environments, where people speak on the phone for a culturally and recreational purposes rich and delightful encounter as a part of living and working.
- An intelligent city features characteristically dynamic communities that foster the communal spirit and neighborhood.
- A smart city values and builds on scenic environmental assets without damaging the environmental, natural and biological systems.
- Open living room in a smart town. The living areas are personalized, alive and environmental stimuli surroundings, as opposed to the inner living room of houses, in which people chat for a sports and heritage objective as an integral part of life and function on a phone.
- A smart town values and develops on beautiful environmental resources without harming environment, natural or biological systems. A smart town has the characteristic of digital communication promoting a neighborhood and the community sense.

- The Smart City has a comprehensive and efficient management system for collection, disposal of waste of municipal, hospital, industrial and hazardous waste.
- The smarting city has an efficient air pollution control system, and the smart city also has a highly effective waste management system.
- A smart city may generate energy efficient, sustainable sources and similar surroundings.

4.9.6 Smart Living (Quality of Life)

- Strongly common interests in the intelligent city.
- Smart cities document the history, culture and environment of the area and commemorate it.
- A smart city is 24 hours a day and seven days a week dynamic downtown.
- A clever city can give women, children and the elderly the security and safety they need.
- A clever city enhances the urban lifestyle. A clever city creates natural and cultural heritage in order to establish a decent quality of life.
- In an intelligent city, the large picture of urban life is understood, but also little details.
- A smart city has available and transparent public places of excellent quality.
- A clever city has good public infrastructure and amenities
- A clever city is the best environment for women, children and elderly people to live in.
- In an intelligent city, people, life and nature are celebrated in the city during festivals.
- An intelligent city has a ritual (or more) event that embodies the community's ideals and objectives.
- In an in-service city the art, culture, and nature of the city is celebrated and promoted. A in-service city involves artists in improving and enriching the esthetic of the city's daily life.

4.10 Challenges for India's Smart Cities

The Mission for Smart Cities also has its own challenges to face.

- **To make it smart, retrofit existing legacy city infrastructure:** There are several hidden challenges to consider as you study a smart city strategy. The most crucial thing is to recognize the poor areas of the present region, such as 100% availability of water and sanitation, which need the utmost attention.
- **Smart Cities Funding:** The High-Power Expert Committee (HPEC) evaluated an Rs 43,386 per capita investment expense (PCIC) for the 20-year term on urban infrastructure investment figures. This means an annual demand of Rs 35,000 crore. How will these ventures be financed is quite curious when most project needs go into full private investment or PPPs? (Public-private partnership).
- **Three-Tier Governance:** Effective deployment of smart city strategies requires effective horizontal and vertical collaboration between various public utility providers as well as efficient coordination between national (MoUD) and state and local government departments on various issues connected with the financing and common best practices and operational efficiency mechanisms.
- **Provision of clearances in a timely way:** Because everybody in our country is aware of the magnitude of corruption, it could be an important problem. Both approval protocols are available electronically to finish the project on a timely basis and can be cleared in time. **Master plan or city development plan availability:** Many towns in our community do not have master plans or a city development plan that is important in the preparation and implementation of intelligent communities and encapsulates all the needs of a town for more opportunity for its residents.
- **ULBs' Technological Constraints:** Because of the restricted recruiting over a range of years, most ULBs have limited technical tools to guarantee timely and cost-efficiency, service and maintenance in future, along with ULBs' inability to attract the best talents at fair current prices.
- **Program Capacity Building:** Capacity building for 100 smart cities is not a simple job, and the most ambitious projects are being retarded due to shortage of quality workers at the central and national levels. Only under 5% of the central budget will be dedicated to training, contextual, metadata

and a rich financing database sustainable development initiative.

- **Utility services' reliability:** The concentrate is on the reliability of power, water, mobile, or internet networks in every smart city in the world. Smart cities should have universal 24/7 access to electricity; the country's existing supply and distribution infrastructure makes it difficult. Cities need to transition to renewables and rely on green buildings and green transport in order to reduce the need for electricity.

4.11 The Government Should Focus on the Following Main Areas for the Country's Creation of Smart Cities

- **Streets of high quality and public spaces:** Well-planed urban streets and areas contribute to local economy development, mobility, culture, creativity and the future. A strong road scheme operates well at least half of the area that can be used for public space for vehicles, public transit and pedestrian and biker transport; 30% is devoted to highways and 20% is for squares, parks, green spaces and well-connected grids.
- **Mixed urban uses and specialization in restricted land-use:** Different land use planning helps create urban jobs, improve local economy, mitigate dependency and ride on vehicles, facilitate bike, bicycle and other non-motorized transportation, decrease the fragmentation of environments and greenhouse gas emissions, provide public facilities, promote mixed neighborhoods and local markets, encourage healthier cities and build ideal neighborhoods.
- **Connectivity:** The aim underlying connection extension is to expand access to jobs and resources for all and to boost local economy. It encourages the accessibility of fly, public transit, and ICT.
- **Waste Management:** Waste storage modeling and ongoing energy supply.
- **Mixed structure of culture:** The idea seeks to promote cooperation and involvement between different social classes in

the same society and guarantee the access of different forms of housing to equitable urban opportunities.

- **Resilience to urbanism:** RESILITY requires emergency preparedness measures, disaster preparedness strategies, processes, plans, and designs to better respond and mitigate pollution from climate change.
- **Efficiency of Energy and Resources:** This encompasses the planning of demand and resource deterioration through strategic planning, building design, infrastructure and appliance, agriculture, production and services policies and initiatives.
- **Energy networks or smart grids:** Market management, electronic car finance, energy efficiency, and clean energy integration programmers.
- **Norms and laws which are practical and enforceable:** Policy, methods, norms and expectations must be developed that change the fundamental needs of municipal governments in order to adapt to cities' rapid urbanization. With guidelines, standards and legislation should be a participatory approach based on the ideals of fairness and social stability.

4.12 Conclusion

A smart DSS for smarter town has been produced, developed by combining the AHP model with the IF description, as a development of the device thought model. The smart DSS also offers the ability to assess the uniformity of IF principles by creating and handling decisions templates and instances by many users as a shared system. The proposed framework is explicitly designed to facilitate decision making in an atmosphere of Smart Cities.

There are many major challenges facing urban governance in India, such as unplanted development, insufficient healthcare services and a lack of basic in urban and suburban areas, a lack of infrastructure, inadequate transport facilities, and containment of traffic, low electricity supplies, and a lack of basic facilities. It, therefore, takes an hour to build and maintain intelligent cities to solve these problems.

References

1. Abella, A., Ortiz-de-Urbina-Criado, M., De-Pablos-Heredero, C., A model for the analysis of data-driven innovation and value generation in smart cities' ecosystems. *Cities*, Portal Komunikacji Naukowej, 64, C, 47–53, 2017.
2. Albino, V., Berardi, U., Dangelico, M., Smart cities: Definitions, dimensions, performance, and initiatives. *J. Urban Technol.*, 22, 1, 3–21, 2015.
3. AlHogail, A., Improving IoT technology adoption through improving consumer trust. *Technologies*, 6, 64, 1–17, 2018.
4. Althobaiti, A. and Abdullah, M., Medium access control potocols for wireless sensor networks classifications and cross-layering. *Proc. Comput. Sci.*, 65, 4–16, 2015.
5. Alvi, A., Bouk, S., Ahmed, S., Yaqub, M., Sarkar, M., Song, H., BEST-MAC: Bitmap assisted efficient and scalable TDMA-based WSN MAC protocol for smart cities. *IEEE Access.*, 4, 312–322, 2016.
6. Bacic, Z., Jogun, T., Majic, I., Integrated sensor systems for smart cities. *Tech. Gazette*, 25, 1, 277–284, 2018.
7. Chichernea, V., The role of collaborative software and decision support systems in the smarter cities. *JISOM*, 5, 1, 44–50, 2011.
8. Druzdzel, M.J. and Flynn, R.R., *Decision Support Systems: Encyclopedia of Library and Information Science*, Second Edition, A. Kent (Ed.), Marcel Dekker, Inc., New York, 2002.
9. Eckhoff, D. and Wagner, I., Privacy in the smart city-applications, technologies, challenges, and solutions. *IEEE Commun. Surv. Tut.*, 20, 1, 489–516, 2018.
10. Farooq, M., Waseem, M., Khairi, A., Mazhar, S., A critical analysis on security concerns of the Internet of Things (IoT). *Int. J. Comput. Appl.*, 111, 7, 1–6, 2015.
11. Fernandes, B., Silva, F., Analide, C., Neves, J., Crowd sensing for urban security in smart cities. *J. Univers. Comput. Sci.*, 24, 3, 302–321, 2018.
12. Hamilton, S., *Maximizing Your ERP System. A Practical Guide for Managers*, McGraw-Hill, British overseas territory of Bermuda, 2003.
13. Jiang, J.-A., Wan, J.-J., Zheng, X.-Y., Chen, C.-P., Lee, C.-H., Su, L.-K., Huang, W.-C., A novel weather information-based optimization algorithm for thermal sensor placement in smart grid. *IEEE Trans. Smart Grid*, 9, 2, 911–922, 2018.
14. Judge, M., Manzoor, A., Ahmed, F., Kazmi, S., Monitoring of power transmission lines through wireless sensor networks in smart grid, in: *Advances in Intelligent System and Computing*, N. Shakhovska, and M. Medykovskyy, (Eds.), pp. 162–170, Springer, USA, 2017.
15. Kaswan, K.S., Dhatterwal, J.S., Gaur, N.K., Smart grid using internet of things, in: *Integration and Implementation of the Internet of Things through Cloud Computing*, pp. 251–271, IGI Global, Galgotia University, Greater Noida, India, KL University AP, India, 2021.

16. Kaswan, K.S., Dhatterwal, J.S., Kumar, K., Blockchain of internet of things-based earthquake alarming system in smart cities, in: *Integration and Implementation of the Internet of Things through Cloud Computing*, pp. 272–287, IGI Global, Galgotia University, Greater Noida, India, KL University AP, India, 2021.
17. Kaswan, K.S. and Dhatterwal, J.S., The use of machine learning for sustainable and resilient building, in: *Digital Cities Roadmap: IoT-Based Architecture and Sustainable Buildings*, Scrivener Publishing Press, ISBN: 9781119791591, Galgotia University, Greater Noida, India, KL University AP, India, 2021.
18. Kumar, S., Mahapatra, B., Kumar, R., Turuk, K., Security and privacy solution for IRFID based smart infrastructure health monitoring. *2018 Technologies for Smart-City Energy Security and Power (ICSESP)*, 2018.
19. Larik, R., Mustafa, M., Qazi, S., Smart grid technologies in power systems: An overview. *Res. J. Appl. Sci. Eng. Technol.*, 11, 3, 633–638, 2015.
20. Li, W., Song, H., Zeng, F., Policy-based secure and trustworthy sensing for internet of things in smart cities. *IEEE Internet Things J.*, 5, 2, 716–723, 2018.
21. Li, Y., Lin, Y., Geertman, S., The development of smart cities in China. *Proceedings of the 14 Int. Conference on Computers in Urban Planning and Urban Management*, Cambridge, MA, USA, pp. 1–20, 2015.
22. Hanna, M.M., Ahuja, R.K., Winston, W.L., *Developing Spreadsheet-Based Decision Support Systems, Using Excel and VBA for Excel*, www.scribd.com/../Developing-Spreadsheet-Based-Decision-Support-Systems5, *Sage Journal*, Bogazici University, Istanbul, Turkey, 2013. https://doi.org/10.1177/0037549712470169
23. Srinivasan, S., Singh, J., Kumar, V., Multi-agent-based decision support system using data mining and case based reasoning. *IJCSI*, 8, 4, 2, July 2011.
24. Singh, H. and Kaswan, K.S., Clinical decision support systems for heart disease using data mining approach. *IJCSSE*, 5, 2, 19, 2016.

5

Evaluating Solutions to Overcome Blockchain Technology Implementation in Smart City Using a Hybrid Fuzzy SWARA-Fuzzy WASPAS Approach

Shivam Goyal, Vernika Agarwal* **and Sanskriti Goel**

Amity International Business School, Amity University, Noida, Uttar Pradesh, India

Abstract

The enhancement in the technologies and the need to include social sustainability is driving the economies to develop smart cities. The research and development departments have introduced various technologies like centralized databases, cloud storage to make way for the conversion of a city into a smart city. "Blockchain Technology" also known as "Digital Ledger" is the building block for these smart cities. The demand for Blockchain technology is increasing at a very rapid rate and in all sectors and markets. The present study aims in understanding these challenges which hinder the adoption of Blockchain technology in smart cities and provide valuable solutions in solving the problem. The study carries out fuzzy step-wise weight assessment ratio analysis (SWARA) and fuzzy weighted aggregated sum product assessment (WASPAS) by determining the weights of each challenge and its importance and providing a rank to different solutions, which will help in providing solution to the challenges. The study is validated by taking Indian cities into consideration.

Keywords: Smart cities, blockchain, challenges, solution, fuzzy logic

**Corresponding author*: vernika.agarwal@gmail.com

Loveleen Gaur, Vernika Agarwal and Prasenjit Chatterjee (eds.) Decision Support Systems for Smart City Applications, (77–98) © 2023 Scrivener Publishing LLC

5.1 Introduction

The past few decades have witnessed an exponential rise in the world's population that lives in urban areas. Nowadays, more than 55% of the planet's population lives in urban parts and according to the report by the United Nations in the coming 30 years, this rate is predicted to reach 70%, as by 2050, a further 25 billion people are predicted to reside in urban areas. The rapid growth within the world's population in addition to the rapid urbanization process brings forth numerous social, technical, organizational and economic problems, which tend to pose a threat to the environmental and economical sustainability of cities. Hence, the development authorities around the world are actively curious about adopting "smart" concepts to optimize the employment of both tangible and intangible assets [1]. In this regard, the concept of "smart city" is proposed to use the latest information and communication technology (ICT) in an intelligent manner aimed to create a sustainable urban environment and improve the quality of living. The smart city contains a huge range of applications in modern societies like smart building for managing the temperature and lighting system; smart energy for optimizing energy consumption using digital technologies; smart healthcare to push diagnostics; smart technology to enable edge processing and intelligent network connectivity; smart education to facilitate the education system using modern technologies; smart governance to supply digital services and policies from the government; smart security to scale back security risks and protect properties, people and knowledge [2]. In contrast to the standard methods, blockchain technology (that was originally designed for Bitcoin cryptocurrency) facilitates the transfer of digital assets among peers with no intermediaries. Blockchain could be a decentralized, publicly available and immutable shared database that would revolutionize the way peers automate payments, interact, trace and track transactions by completely eliminating the necessity for a central authority for governing the transactions. In traditional systems, the data collected is stored on a central server for future preference. The servers are prone to several unwanted outcomes like revealing sensitive information due to hacking and data theft. This brings the requirement of shifting our interest from the traditional concept of data storage to the modern concept i.e., Blockchain technology. Blockchain, a comparatively recent technological trend, maybe a distributed network and chain of cryptographic blocks combined together to create a peer-to-peer (P2P) network is decentralized and distributed in nature. Within the blockchain, each node has its own distributed ledger for storing the history of the transactions [3]. Blockchain is often leveraged to realize authentication, authorization, accountability, security, integrity, confidentiality and non-repudiation for real-time applications, which can never be provided by a centralized system efficiently in an exceedingly smart community setting.

The main focus of the study is the validation of the following objectives:

- To pen down the challenges occurring in the implementation of blockchain technology in smart cities and suggesting the solutions which can be carried out to overcome the challenges.
- To compute the weights for each challenge and provide a rank to different solutions which would help in overcoming the challenges by utilizing Fuzzy SWARA and Fuzzy WASPAS.

The following paper helps in finding out the challenges occurring in the implementation of blockchain technology in smart cities and suggesting the solutions which can be carried out to overcome the challenges. The evaluation of the shortlisted challenges and the solution is done by utilizing Fuzzy SWARA and Fuzzy WASPAS.

The research is classified into the following sections, as in Section 5.1, the context of the formulated aim is briefed. Section 5.2 elaborates the research methodology. Section 5.3 discussion of research design, followed by numerical illustration in Section 5.4. Section 5.5 discusses the conclusion and limitations of the study.

5.2 Research Methodology

The detailed steps are given in subsequent sub-section followed by the flowchart of the methodology in Figure 5.1.

5.2.1 Stepwise Weight Assessment Ratio Analysis (SWARA) Method

Step 1: The criteria's finalized are arranged on the basis of their importance. Starting from the most importance to the least importance.

To evaluate the importance of the shortlisted challenges linguistic scale is used as represented in (Table 5.1). Using the Linguistic Scale the level of uncertainty is reduced.

Step 2: The process for the methodology begins with the second criterion, where the linguistic values are provided to each criteria j, which is based on the relative importance of the early $(j-1)$ criterion. The ratio obtained in this step is known as Comparative Importance of the Average Value.

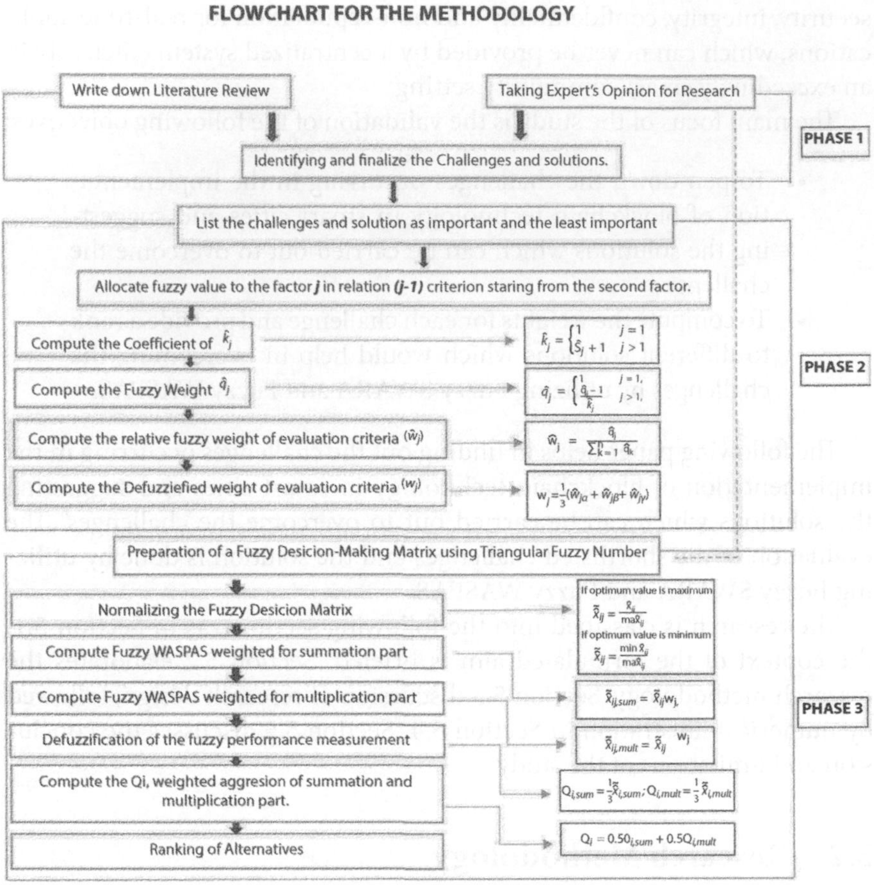

Figure 5.1 Flowchart of the methodology.

Table 5.1 Fuzzy evaluation scale.

Linguistic variable	Fuzzy scale
Extremely unimportant (A)	(0.0,0.0,0.1)
Not very important (B)	(0.0,0.1,0.3)
Not important (C)	(0.1,0.3,0.5)
Fair (D)	(0.3,0.5,0.7)
Important (E)	(0.5,0.7,0.9)
Very Important (F)	(0.7,0.9,1)

Step 3: In this step the Fuzzy coefficient k_j is determined using:

$$k_j = \begin{cases} 1 & j = 1 \\ S_j + 1 & j > 1 \end{cases} \tag{5.1}$$

Step 4: In this step recalculated weight q_j is determined using:

$$q_j = \begin{cases} 1 & j = 1 \\ q_{(j-1)/k_j} & j > 1 \end{cases} \tag{5.2}$$

Step 5: In this step the computation of the relative fuzzy weights is carried out using:

$$w_j = \frac{q_j}{\sum_{k=1}^{n} q_k} \tag{5.3}$$

where:

w_j = Relative weight of the j criterion.
n = Criteria number.

Step 6: In this step, the relative weights obtained in the previous step are defuzziefied using the center of area method.

$$w_j = \frac{1}{3}\hat{w}_j = \frac{1}{3}(\hat{w}_{j\alpha} + \hat{w}_{j\beta} + \hat{w}_{j\gamma}) \tag{5.4}$$

5.2.2 Weighted Aggregated Sum Product Assessment (WASPAS) Method

WASPAS relatively belongs to the family of MCDM techniques which has a unique and different combination of outcomes of two different models i.e., Weighted sum model (WSM) and Weighted product model (WPM).

The Prioritization of the alternatives is carried forward on the basis of combined optimality criteria value which is calculated from the outcomes of the above-mentioned Models. The criteria weights which have to be used can either be obtained from some expert or by help of a particular method. In our study the weights are obtained by the help of Fuzzy SWARA methodology. By doing the sensitive analysis within its own function, Fuzzy WASPAS can check the consistency of alternative rankings.

Step 1: In the first step the preparation for the fuzzy decision-making matrix is carried out using the triangular fuzzy number as shown below:

$$X = \begin{bmatrix} x_{11} & x_{12} & \cdots & x_{im} \\ x_{21} & x_{22} & \cdots & x_{2m} \\ \cdots & \cdots & \cdots & \cdots \\ x_{n1} & x_{n2} & \cdots & x_{mn} \end{bmatrix} \tag{5.5}$$

where:

m = numbers of evaluation alternatives,

n = number of evaluation criteria's (factors)

xij = fuzzy evaluation of the ***ith*** alternative with respect to ***jth*** criterion.

Step 2: This step involves the Normalization of the fuzzy decision-matrix: after the normalization is performed the element identified is termed as $\hat{x}_{ij}$.

If optimum value achieved is maximum then:

$$\tilde{x}_{ij} = \frac{\mathrm{x_{ij}}}{\max \mathrm{x_{ij}}} \; j = 1, \ldots\ldots, n; \mathrm{i} = 1, \ldots, m \tag{5.6}$$

If optimum value achieved is minimum then:

$$\tilde{x}_{ij} = \frac{\mathrm{x_{ij}}}{\min \mathrm{x_{ij}}} \; j = 1, \ldots\ldots, n; \mathrm{i} = 1, \ldots, m \tag{5.7}$$

where $\hat{x}_{ij}$ = normalized value of $\hat{x}_{ij}$

Step 3: In the step the Computation for the Fuzzy WASPAS normalized decision-making weighted matrix for the summation part is carried out using:

$$\tilde{\hat{\mathrm{x}}}_{ij,\ sum} = \tilde{\hat{x}}_{ij} \mathrm{w_j}, \quad j = 1, \ldots, n: \quad i = 1, \ldots m$$

$$\tilde{\hat{\mathrm{X}}}_{i,\ sum} = \sum_{j=1}^{n} \tilde{\hat{\mathrm{x}}}_{ij,\ sum} \tag{5.8}$$

where:

Wj = weight (relative importance) of significance (weight) of ***jth*** criterion.

Step 4: In the step the Computation for the Fuzzy WASPAS normalized decision-making weighted matrix for the multiplication part is carried out using:

$$\hat{\tilde{X}}_{ij,\ mult} = \hat{\tilde{x}}_{ij}^{\ w_j}, \quad j=1,\ldots,n: \quad i=1,\ldots m$$

$$\hat{\tilde{X}}_{i,\ mult} = \prod_{j=1}^{n} \hat{\tilde{x}}_{ij,\ smult} \tag{5.9}$$

Step 5: In this step, the defuzzification for the fuzzy performance measurement is carried out using center-of-area method, which is considered as the most practical and easy way.

$$Q_{i,sim} = \frac{1}{3}\hat{\tilde{X}}_{i,\ sum} = \frac{1}{3}\left(\hat{\tilde{X}}_{i,\ sum,\ \alpha} + \hat{\tilde{X}}_{i,\ sum,\beta} + \hat{\tilde{X}}_{i,\ sum,\gamma}\right)$$

$$Q_{i,mult} = \frac{1}{3}\hat{\tilde{X}}_{i,\ mult} = \frac{1}{3}\left(\hat{\tilde{X}}_{i,\ mult,\ \alpha} + \hat{\tilde{X}}_{i,\ mult,\beta} + \hat{\tilde{X}}_{i,\ mult,\gamma}\right) \tag{5.10}$$

Step 6: In this step, the computation of Qi is carried out i.e., weighted aggregation of the multiplication and summation part.

$$Q_i = 0.5Q_i^{(1)} + 0.5Q_i^{(2)} = 0.5\sum_{j=1}^{n} \tilde{x}_{ij} w_j + 0.5\sum_{j=1}^{n} (\tilde{x}_{ij})^{w_j} \tag{5.11}$$

Step 7: In this step, the determination of the total relative significance of ***ith*** alternative a more generalized equation is used to enhance the ranking and accuracy effectiveness

$$Q_i = \lambda Q_i^{(1)} + \lambda Q_i^{(2)} = \lambda\sum_{j=1}^{n} x_{ij} w_j + \lambda\sum_{j=1}^{n} (x_{ij})^{w_j} \tag{5.12}$$

All the given alternatives can be ranked as per the Q_i values i.e., the top alternative having the highest Q_i value would be the best. λ represents the parameters of the Fuzzy WASPAS and can be changed in between the range of 0–1.

λ = 0, i.e., fuzzy WASPAS leads to WPM.

λ = 1, i.e., WASPAS transformed to WSM.

5.3 Research Design

The main motive behind the research design is to identify the challenging factors, which can fulfill the stakeholder requirement by evaluating solutions to overcome blockchain technology Implementation in smart city using a hybrid methodology. An extensive literature survey and interaction with different stakeholders was carried out to list down the list of challenging factors. Several stakeholders were identified for the interpretation of the data. This assorted team of stakeholders were selected to incorporate various perspectives into the decision-making process. Based on the consensus of these stakeholders, the following challenges and solutions were shortlisted (Table 5.2 and Table 5.3).

Table 5.2 List of shortlisted challenges.

S. no.	Challenges	Explanation	References
C1.	Cryptographic vulnerability & Rules Violation	It refers to the threats to encrypted data due to cyber censure from unauthorized sources by violating all the rules.	[4]
C2.	Centralized system of Connecting devices	Centralized system refers to the concentration of client's data at one destination i.e., central user. It is the most customary system where the request placed by the connecting users is processed by the central user.	[4]
C3.	Scalability and Performance	The measurement of productivity and further assessment refers to scalability and performance. Higher the scalability, better the performance.	[4]

(*Continued*)

Table 5.2 List of shortlisted challenges. (*Continued*)

S. no.	Challenges	Explanation	References
C4.	Costing	Costing as a challenge in blockchain technology refers to the expenses incurred in procurement of technology and further application in the advancement of smart cities.	[1]
C5.	Energy Usage in Each Plot of city	The consumption of energy every day in running errands is different in every part of the city because of different energy requirements in different parts of the city.	[5]
C6.	Data Tracking & Management	Keeping the trace, managing and backing up the data packets refers to data tracking and management. It is important to keep the data backup for further usage purposes.	[5]
C7.	Traffic congestion	Number of users accessing the server at the same time results in slowing down the processes and hence leading to Traffic congestion.	[6]
C8.	Translucency in governmental procedures	The legal policies execution of our country is very slow and time taking, due to which it takes years for a single policy implementation hence leading to translucency in government procedures.	[6]
C9.	Unified implementation	Application of varied functions on similar grounds in all the smart cities across and similar processes being carried out refers to Unified implementation.	[6]

Table 5.3 List of shortlisted solutions.

S. no.	Solutions	Explanation	References
S1.	Stronger Encryption	Use of best possible security and latest technology to secure and encrypt all the data present on servers resulting in Stronger Encryption.	[7]
S2.	Distributed nodes of data packets	Distributed nodes of data packets means that there should be sub-servers connected to the central server with the help of nodes so that the traffic gets distributed.	[8]
S3.	Byzantine fault-tolerant algorithms	The methodology of distributed networks to come upon a consensus even in case of failure in the nodes of a network in order to somehow complete the task is derived from Byzantine fault-tolerant algorithms.	[9]
S4.	Cost effective methodology	It means procurement of cost-effective technology and best possible utilization of same with utmost effectiveness and efficiency.	[10]
S5.	Energy usage on eco mode in each plot of the city	The energy used should be sustained and recycled for future use in order to bring out an Eco-friendly environment.	[7]

(*Continued*)

Table 5.3 List of shortlisted solutions. (*Continued*)

S. no.	Solutions	Explanation	References
S6.	Real-time data monitoring and backup arrangement	The data available on the servers should be monitored and evaluated on a real-time basis and a proper backup system must be installed in case of loss of data.	[11]
S7.	Smart transportability	The systems must have automated analyzing methodology in order to avoid traffic congestion and hence maintain Smart Transportability.	[12]
S8.	Transparency in the government essential procedure	There must be a specialized department to look into the grievances and problems and must solve them at the earliest possible.	[6]
S9.	Decentralized implementation	The implementation of blockchain technology should be based on decentralized methodology and should be real time based in order to keep each level of authority updated.	[6]

5.4 Application of Proposed Methodology

This section of the paper discusses the methodology used in the paper. After discussion with experts and different research publications, a finalized list of the challenges was prepared which hindered the implementation of blockchain technology in smart cities and proposing solutions according to the challenges, the analysis of the data was carried out. Fuzzy SWARA methodology was used in the initial stage. After brainstorming with all the listed challenges, the challenges were arranged in the form of most important to least important on the basis of Fuzzy evaluation

Table 5.4 Results of fuzzy SWARA to weight challenges.

	Comparative importance (most important to least important)	Coefficient $kj = Sj + 1$	Recalculated fuzzy weight $Qj = Q(j\text{-}1)/Kj$			Relative weight $w_j = \frac{\hat{q}_j}{\sum_{k=1}^{n} \hat{q}_k}$			Defuzzified relative weight Wj
C6		1.0 1.0 1.0	1.000	1.000	1.000	0.222	0.398	0.419	0.760
C2	0.7 0.9 1	1.7 1.9 2.0	0.588	0.526	0.500	0.131	0.209	0.209	0.410
C8	0.7 0.9 1	1.7 1.9 2.0	0.345	0.276	0.250	0.077	0.110	0.105	0.221
C1	0.7 0.9 1	1.7 1.9 2.0	0.791	0.145	0.125	0.176	0.058	0.052	0.251
C9	0.5 0.7 0.9	1.5 1.7 0.9	0.527	0.085	0.138	0.117	0.034	0.058	0.170
C4	0.3 0.5 0.7	1.5 1.7 0.9	0.351	0.167	0.153	0.078	0.066	0.064	0.166
C7	0.1 0.3 0.5	1.1 1.3 1.5	0.319	0.128	0.102	0.071	0.051	0.043	0.136
C3	0.1 0.3 0.5	1.1 1.3 1.5	0.290	0.098	0.068	0.064	0.039	0.028	0.113
C5	0.0 0.1 0.3	1.0 1.1 1.3	0.290	0.089	0.052	0.064	0.035	0.022	0.107
		SUM =	**4.501**	**2.514**	**2.388**				

Table 5.5 Fuzzy waspas decision making matrix.

WEIGHTS		0.251			0.410			0.113			0.166			0.107			0.760			0.136			0.221			0.170		
		C1			*C2*			*C3*			*C4*			*C5*			*C6*			*C7*			*C8*			C9		
	Challenges ➡	*Cryptographic vulnerability & rules violation*			*Centralized system of connecting devices*			*Scalability and perfor-mance*			*Costing*			*Energy use in each plot of city*			*Data tracking manage-ment*			*Traffic congestion*			*Translucency in government procedures*			*Unified implementation*		
	Solutions ⬇																												
S1	Stronger Encryption	0.9	1.0	1.0	0.9	1.0	1.0	0.3	0.5	0.7	0.3	0.5	0.7	0.0	0.0	0.1	0.9	1.0	1.0	0.5	0.7	0.9	0.0	0.0	0.1	0.5	0.7	0.9
S2	Distributed nodes of data packets	0.0	0.0	0.1	0.9	1.0	1.0	0.7	0.9	1.0	0.0	0.1	0.3	0.3	0.5	0.7	0.9	1.0	1.0	0.7	0.9	1.0	0.0	0.0	0.1	0.7	0.9	1.0
S3	Byzantine fault-tolerant algorithms	0.3	0.5	0.7	0.3	0.5	0.7	0.9	1.0	1.0	0.0	0.1	0.3	0.0	0.0	0.1	0.5	0.7	0.9	0.0	0.0	0.1	0.0	0.1	0.3	0.0	0.1	0.3
S4	Cost effective methodology	0.0	0.1	0.3	0.5	0.7	0.9	0.0	0.1	0.3	0.9	1.0	1.0	0.5	0.7	0.9	0.3	0.5	0.7	0.9	1.0	1.0	0.9	1.0	1.0	0.7	0.9	1.0
S5	Energy usage on Eco mode in Each Plot Of city	0.0	0.0	0.1	0.0	0.1	0.3	0.0	0.1	0.3	0.3	0.5	0.7	0.9	1.0	1.0	0.0	0.0	0.1	0.3	0.5	0.7	0.0	0.0	0.1	0.0	0.0	0.1

(Continued)

Table 5.5 Fuzzy waspas decision making matrix. (*Continued*)

	WEIGHTS	*0.251*			*0.410*			*0.113*			*0.166*			*0.107*			*0.760*			*0.136*			*0.221*			*0.170*		
		C1			*C2*			*C3*			*C4*			*C5*			*C6*			*C7*			*C8*			*C9*		
	Challenges ➡	*Cryptographic vulnerability & rules violation*			*Centralized system of connecting devices*			*Scalability and perfor-mance*			*Costing*			*Energy use in each plot of city*			*Data tracking manage-ment*			*Traffic congestion*			*Translucency in government procedures*			*Unified implementation*		
	Solutions ⬇																											
S6	**Real time data monitoring & backup arrangement**	0.5	0.7	0.9	0.3	0.5	0.7	0.3	0.5	0.7	0.0	0.1	0.3	0.3	0.5	0.7	0.9	1.0	1.0	0.1	0.3	0.5	0.0	0.1	0.3	0.0	0.0	0.1
S7	**Smart Transportability**	0.0	0.0	0.1	0.5	0.7	0.9	0.0	0.0	0.1	0.0	0.0	0.1	0.0	0.1	0.3	0.7	0.9	1.0	0.9	1.0	1.0	0.3	0.5	0.7	0.3	0.5	0.7
S8	**Transparency in the government essential procedure**	0.9	1.0	1.0	0.7	0.9	1.0	0.0	0.0	0.1	0.0	0.0	0.1	0.0	0.1	0.3	0.0	0.1	0.3	0.0	0.1	0.3	0.9	1.0	1.0	0.9	1.0	1.0
S9	**Decentralized implementation**	0.7	0.9	1.0	0.5	0.7	0.9	0.3	0.5	0.7	0.7	0.9	1.0	0.7	0.9	1.0	0.3	0.5	0.7	0.7	0.9	1.0	0.9	1.0	1.0	0.9	1.0	1.0

Table 5.6 FUZZY WASPAS normalized decision-making matrix.

Challenges ➡	*C1*			*C2*			...			...			*C8*			*C9*		
Solutions ⬇																		
S1	0.900	1.000	1.000	0.900	1.000	1.000	...	...	...	...	...	...	0.000	0.000	0.100	0.500	0.700	0.900
S2	0.000	0.000	0.100	0.900	1.000	1.000	...	...	...	...	...	...	0.000	0.000	0.100	0.700	0.900	1.000
S3	0.300	0.500	0.700	0.300	0.500	0.700	...	...	...	...	...	...	0.000	0.100	0.300	0.000	0.100	0.300
S4	0.000	0.100	0.300	0.500	0.700	0.900	...	...	...	...	...	...	0.900	1.000	1.000	0.700	0.900	1.000
S5	0.000	0.000	0.100	0.000	0.100	0.300	...	...	...	...	...	...	0.000	0.000	0.100	0.000	0.000	0.100
S6	0.500	0.700	0.900	0.300	0.500	0.700	...	...	...	...	...	...	0.000	0.100	0.300	0.000	0.000	0.100
S7	0.000	0.000	0.100	0.500	0.700	0.900	...	...	...	...	...	...	0.300	0.500	0.700	0.300	0.500	0.700
S8	0.900	1.000	1.000	0.700	0.900	1.000	...	..	...	...	...	...	0.900	1.000	1.000	0.900	1.000	1.000
S9	0.700	0.900	1.000	0.500	0.700	0.900	...	...	...	...	...	...	0.900	1.000	1.000	0.900	1.000	1.000

Table 5.7 Fuzzy WASPAS weighted normalized decision-making matrix for summation part.

Challenges ➡	**C1**			**C2**			**...**			**...**			**C8**			**C9**		
Solutions ⬇																		
S1	0.226	0.251	0.251	0.369	0.410	0.369	...	...	...	...	...	...	0.000	0.000	0.022	0.085	0.119	0.153
S2	0.000	0.000	0.025	0.369	0.410	0.410	...	...	...	...	...	...	0.000	0.000	0.022	0.119	0.153	0.170
S3	0.075	0.126	0.176	0.123	0.205	0.410	...	...	...	...	...	...	0.000	0.022	0.066	0.000	0.017	0.051
S4	0.000	0.025	0.075	0.205	0.287	0.287	...	...	...	...	...	...	0.153	0.221	0.221	0.119	0.153	0.170
S5	0.000	0.000	0.025	0.000	0.041	0.369	...	...	...	...	...	...	0.000	0.000	0.022	0.000	0.000	0.017
S6	0.126	0.176	0.226	0.123	0.205	0.123	...	...	...	...	...	...	0.000	0.022	0.066	0.000	0.000	0.017
S7	0.000	0.000	0.025	0.205	0.287	0.287	...	...	...	...	...	...	0.051	0.111	0.155	0.051	0.085	0.119
S8	0.226	0.251	0.251	0.287	0.369	0.369	...	...	...	...	...	...	0.153	0.221	0.221	0.153	0.170	0.170
S9	0.176	0.226	0.251	0.205	0.287	0.410	...	...	...	...	...	...	0.153	0.221	0.221	0.153	0.170	0.170

Table 5.8 Fuzzy WASPAS weighted normalized decision-making matrix for multiplication part.

Challenges ➡	C1			C2			…			…			C8			C9		
Solutions	⬇																	
S1	0.974	1.000	1.000	0.958	1.000	1.000	…	…	…	…	…	…	0.000	0.000	0.601	0.889	0.941	0.982
S2	0.000	0.000	0.561	0.958	1.000	1.000	…	…	…	…	…	…	0.000	0.000	0.601	0.941	0.982	1.000
S3	0.739	0.840	0.914	0.610	0.753	0.864	…	…	…	…	…	…	0.000	0.601	0.766	0.000	0.676	0.815
S4	0.000	0.561	0.739	0.753	0.864	0.958	…	…	…	…	…	…	0.977	1.000	1.000	0.941	0.982	1.000
S5	0.000	0.000	0.000	0.000	0.389	0.610	…	…	…	…	…	…	0.000	0.000	0.601	0.000	0.000	0.676
S6	0.840	0.914	0.974	0.610	0.753	0.864	…	…	…	…	…	…	0.000	0.601	0.766	0.000	0.000	0.676
S7	0.000	0.000	0.561	0.753	0.864	0.958	…	…	…	…	…	…	0.766	0.858	0.924	0.815	0.889	0.941
S8	0.974	1.000	1.000	0.864	0.958	1.000	…	…	…	…	…	…	0.977	1.000	1.000	0.982	1.000	1.000
S9	0.914	0.974	1.000	0.753	0.864	0.958	…	…	…	…	…	…	0.977	1.000	1.000	0.982	1.000	1.000

Table 5.9 Fuzzy WASPAS result and ranking of the solutions.

	Aggregate fuzzy summation value			Q_i (SUM)	Q_i/5 (SUM)	Aggregate fuzzy multiplication value			Q_i (MULT)	Q_i/5 (MULT)	ΣQi*5	Ranking
S1	1.516	1.775	1.867	1.7192	0.3438	0.000	0.000	0.412	0.1373	0.2746	3.0922	3
S2	1.396	1.647	1.780	1.6076	0.3215	0.000	0.000	0.266	0.0886	0.1772	2.4938	4
S3	0.680	1.009	1.514	1.0676	0.2135	0.000	0.000	0.213	0.0710	0.1421	1.7779	7
S4	0.907	1.275	1.550	1.2437	0.2487	0.000	0.209	0.466	0.2248	0.4497	3.4921	2
S5	0.240	0.369	0.898	0.5026	0.1005	0.000	0.000	0.000	0.0000	0.0000	0.5026	9
S6	1.030	1.338	1.439	1.2688	0.2538	0.000	0.000	0.300	0.1001	0.2002	2.2700	5
S7	0.910	1.209	1.405	1.1746	0.2349	0.000	0.000	0.216	0.0720	0.1441	1.8950	6
S8	0.666	0.896	1.137	0.8995	0.1799	0.000	0.000	0.157	0.0524	0.1048	1.4237	8
S9	1.123	1.541	1.910	1.5246	0.3049	0.140	0.440	0.701	0.4270	0.8540	5.7948	1

scale (Table 5.1). After the arrangement of the challenges Eq. (5.1) is used to carry out the calculation for the fuzzy coefficient ***kj***. Subsequently Eq. (5.2) and Eq. (5.3) used to calculate the Recalculated Fuzzy Weight ***Qj*** and the Relative fuzzy Weight ***Wj*** of the listed challenges. After this Defuzzification of Fuzzy Relative Weight ***Wj*** is carried out for each of the challenges using the Eq. (5.4), respectively (Table 5.4).

After the calculation of Weights for the challenges with the help of Fuzzy SWARA, Fuzzy WASPAS is applied to rank the solution to overcome the challenges faced in the implementation of the Blockchain Technology. The solutions were evaluated on the basis of fuzzy scale (Table 5.1) and Fuzzy WASPAS decision matrix is obtained (Table 5.5).

In the next step Eq. (5.6) and Eq. (5.7) is used to calculate Fuzzy WASPAS Normalized Decision Matrix (Table 5.6).

After the normalized decision matrix is obtained using the, then Eq. (5.8) and Eq. (5.9) are used to compute the fuzzy WASPAS normalized decision-making weighted matrix for the summation part and the multiplication part (Table 5.7 and Table 5.8). Defuzzification of the fuzzy performance measurement is done using the center-of-area method by using Eq. (5.10). Then Eq. (5.11) is applied to compute ***Qi***, the weighted aggregation of the summation and multiplication part. The Solutions were then provided ranks on the basis of the Qi values calculated. Table 5.9 shows the result of the Fuzzy WASPAS.

5.5 Conclusion

The present study is intended to identify and prioritize the solutions to mitigate the impact of challenges in the implementation of blockchain technology in smart cities. The implementation of blockchain technology in smart cities is which every city is aiming to do since it has a positive result as all the required information is present in one place at any time but a number of challenges hinder this implementation. The list challenges shortlisted for the study include all kinds of barriers, i.e., governmental and normal challenges. The study is conducted using a modified fuzzy SWARA and fuzzy WASPAS methodology to evaluate the solutions and challenges and then provide ranks to the solutions to overcome the challenges. The contribution of the study is conducted two-fold, the computation of weights for the challenges using the Fuzzy SWARA. The findings reflect that "Data Tracking & Management (C6)." After obtaining the weights of all the challenges, the ranking to solution is computed using Fuzzy WASPAS. The finding reflects that "decentralized implementation (S9)" is the leading solution to overcome the challenges in blockchain technology implementation in smart cities. All the other solutions are important as well, which can have a lot of impact in overcoming all the shortlisted challenges since the data are focused on collection of data on one-to-one basis with the experts and literature review. In future the research can include a greater number of experts. The data can be collected on a larger scale, and a comparison can be done between different countries. Future research can focus on validating the research with other methodologies like Grey Methodology, DEMATEL, AHP etc. or a combination of two methodology as per the requirements. The future scope of the study can be collection of data on a large scale via questionnaire, which will result in giving outcome in other aspects of study at large scale.

References

1. Aggarwal, S., Chaudhary, R., Aujla, G.S., Kumar, N., Choo, K.K.R., Zomaya, A.Y., Blockchain for smart communities: Applications, challenges and opportunities. *J. Netw. Comput. Appl.*, 144, 13–48, 2019.
2. Prajapati, H., Kant, R., Shankar, R., Prioritizing the solutions of reverse logistics implementation to mitigate its barriers: A hybrid modified SWARA and WASPAS approach. *J. Clean. Prod.*, 240, 118219, 2019.
3. Pieroni, A., Scarpato, N., Di Nunzio, L., Fallucchi, F., Raso, M., Smarter city: Smart energy grid based on blockchain technology. *Int. J. Adv. Sci. Eng. Inf. Technol.*, 8, 1, 298–306, 2018.
4. Bhushan, B., Khamparia, A., Sagayam, K.M., Sharma, S.K., Ahad, M.A., Debnath, N.C., Blockchain for smart cities: A review of architectures, integration trends and future research directions. *Sustain. Cities Soc.*, 61, 102360, 2020.
5. Chaurasia, V.K., Yunus, A., Singh, M., An overview of smart city: Observation, technologies, challenges and blockchain applications, in: *Blockchain Technology for Smart Cities*, pp. 133–154, Springer, Singapore, 2020.
6. Hakak, S., Khan, W.Z., Gilkar, G.A., Imran, M., Guizani, N., Securing smart cities through blockchain technology: Architecture, requirements, and challenges. *IEEE Netw.*, 34, 1, 8–14, 2020.
7. Nam, K., Dutt, C.S., Chathoth, P., Khan, M.S., Blockchain technology for smart city and smart tourism: Latest trends and challenges. *Asia Pac. J. Tour. Res.*, 26, 4, 454–468, 2019.
8. Aujla, G.S., Singh, M., Bose, A., Kumar, N., Han, G., Buyya, R., Blocksdn: Blockchain-as-a-service for software defined networking in smart city applications. *IEEE Network*, 34, 2, 83–91, 2020.

9. Liang, X., Shetty, S., Tosh, D., Exploring the attack surfaces in blockchain enabled smart cities, in: *2018 IEEE International Smart Cities Conference (ISC2)*, IEEE, pp. 1–8, September 2018.
10. Sharma, P.K. and Park, J.H., Blockchain based hybrid network architecture for the smart city. *Future Gener. Comput. Sy.*, 86, 650–655, 2018.
11. Chen, W., Xu, Z., Shi, S., Zhao, Y., Zhao, J., A survey of blockchain applications in different domains, in: *Proceedings of the 2018 International Conference on Blockchain Technology and Application*, pp. 17–21, December 2018.
12. Persaud, P., Varde, A.S., Robila, S., Enhancing autonomous vehicles with commonsense: Smart mobility in smart cities, in: *2017 IEEE 29th International Conference on Tools with Artificial Intelligence (ICTAI)*, IEEE, pp. 1008–1012, 2017.

6

Identification and Analysis of Challenges and Their Solution in Implementation of Decision Support System (DSS) in Smart Cities

Shreya Gupta, Shubhanshi Mittal and Vernika Agarwal*

Amity International Business School, Amity University, Noida, Uttar Pradesh, India

Abstract

Smart city is an intelligent and high-tech city that connects people and things with the help of new technologies to make the city more sustainable. Smart city will become smart by providing its people smart solutions, which can be through decision support system (DSS). DSS can be used in modeling a smart city with the use of Internet of Things (IoT), cloud services, artificial intelligence (AI), everything-as-a-service (XaaS) position it ahead of many other softwares. The leading step has been the identification of challenges faced in implementing the software and understanding the solutions that can be provided to overcome these challenges. The primary purpose of the research is to identify these challenges, which withhold the use of the DSS in smart cities and provide optimal solution to them by using multicriteria decision making (MCDM) approach. The MCDM approach of fuzzy-technique for order of preference by similarity to ideal solution (TOPSIS) has been used. The results have been validated using the case of smart cities situated in the northerly part of India.

Keywords: Smart city, decision support system (DSS), fuzzy TOPSIS, challenges, software

**Corresponding author*: vernika.agarwal@gmail.com

Loveleen Gaur, Vernika Agarwal and Prasenjit Chatterjee (eds.) Decision Support Systems for Smart City Applications, (99–118) © 2023 Scrivener Publishing LLC

6.1 Introduction

Decision support system (DSS) provides the decision maker with a software tool, which is strategic and operational in nature and analyzes big data, which is required to solve problems in a given project. Because of the rapid evolution in information and communication technology (ICT), DSS get a unique advantage since it is continuously evolving. DSS can be used in modeling a smart city as the use of Internet of Things (IoT), cloud services, artificial intelligence (AI), everything-as-a-service (XaaS) position it ahead of many other software.

Smart city has emerged as a favored response for urban population in order to have a quality life. People are moving from villages to cities i.e. from rural areas to urban areas. Nowadays we see that urban population is around 63% of gross domestic product in India. And it is assumed that it will reach to 75% in the next 15 years. In order to deal with the large-scale urbanization, it is important to make plans to handle the urbanization effectively and efficiently. The aim of the urban planners is to develop the urban ecosystem. The strategies used to develop the urban ecosystem is by city improvement, redevelopment, and by doing city extensions.

The smart city consists of recognizing, analyzing, and integrating with the help of information and communication technologies. Key information of the basic system of urban operation in order to give an intelligent answer to various Requirements including people's livelihood, environmental protection, public safety, urban services, and industrial and commercial activities. Its essence is to use advanced information technology to realize the intelligent management and operation of the city to create better life for the people in the city and promote the harmonious and sustainable growth of the city. The mission of smart city cannot be completed if it does not involve smart people. Another strategy for smart city development is a pan city initiative, which indulge smart people to actively participate in deploying smart solutions and implementing reforms and also in using more and more data, information and technology.

A smart city uses decision support system in a way that is adaptable, efficient, scalable, available, secure, safe, and resilient, so that it can improve the citizens' quality of life; ensure concrete financial development for its people, such as greater living norms and job prospects; improve residents' well-being, including medical treatment, healthcare, physical security, and schooling. Establish an eco-accountable and viable strategy that serves today's requirements without sacrificing potential ages' requirements [1].

But use of decision support system (DSS) has some challenges, which are discussed in the paper. After carefully analyzing the various aspects of DSS, 11 challengers were extracted along with their solution to conclude.

Thus, the objectives of the study are:

- To recognize and analyze various challengers faced in implementing DSS (decision support system) in smart cities.
- To find the suitable solution to each challenger.
- To validate the domain model with the help of experts.

DSS reduces the decision-making time as it analyzes data and present it to the right user at the right time. The quality of the result is also better as there is a large amount of data, which are analyzed in order to obtain a result.

In order to handle the large-scale urbanization, the cities must take various steps to improve sustainability. Now, there are a large number of smart solutions that can improve the cities efficiencies.

6.2 Review of Literature

A lot of urban problems have evolved, which affect the standard of living of people in a different way, socially, economically, and environmentally. Smart city is a like a solution to these problems, as it provides quality living [2]. Smart cities have positively transformed the quality of life of its residents. With the help of ICT, smart cities will be developed and hence have been the center of attraction for builders [3]. A major role in developing smart cities is played by various technologies, which have been the biggest enabler [4]. The use of advanced technology and software like IoT, everything-as-a-service (XaaS), ICT, blockchain, provide the decision makers with a plan of action, which is required to carry out the project [5]. IoT, along with blockchain, was used provide better services to the residents of the smart cities, optimally using the resources. The use of interconnected smart objects is beneficial for the communication between the residents and the government [6]. Network technology and communication framework have safeguarded an intracommunication framework, enabling smart objects of the city to connect with each other [7]. Smart object or devices which are interconnected are rooted in the environment. With the help of IoT, these smart devices collect the information which help in formulating the models [8]. The architecture model, due to the broad scope of smart city, is not yet standardized. A feasible architectural model for smart cities is built with the help of DSS and AI [9]. When AI is used with machine learning (ML), a set of data algorithms is formed. These sets of data formulate a smart application for smart city [10]. This kind of application and framework revolves around the concerns related to the urban problems. So, ICT and DSS help to solve these problems [11]. Park *et al.* [12] researched that ICT and IoT form the base to infrastructure development in smart cities. They are the key technology which have enabled the formulation of smart cities. Development of these smart cities, in the future will become the strength of the country's development index. Big data and IoT and their advancement software are the key to this development [13]. Literature has researched the work of various authors and researchers and has found that the formulation of smart cities is not possible without the usage of technology and advance software to analyze and assemble big data. As the focus of various researchers has been on how technology, namely IoT, ICT, blockchain, AI and ML, hence the focus of this study is to understand the usage of DSS with respect to smart cities.

Smart cities continue to improve the standard of living of its residents. With the help of hi-tech devices, software and advance technology, building of smart cities has become easier than anticipated [3]. DSS and its constituents, because of their strategical and operational approach, help the decision makers with the ongoing projects [5]. IoT, being one of the components of DSS, has helped with waste management in smart cities. IoT has enabled the surveillance system, which help in efficient waste collection [14]. Kashevnik and Lashkov [15] demonstrated the possibilities of DSS to drivers and passengers, which will enable them for a more comfortable trip. DSS having the quality to analyze big data, helps to maintain sustainability in smart cities. It analyzes the various sources of CO_2 (carbon dioxide) and proposes ways to reduce the same [16]. This reduction of CO_2, is also associated with the communication between the local residents and government. This communication is done using IoT and various interconnected devices, which is the base of AI [6]. This successful implementation of artificial intelligences makes the DSS more intelligent [17] and has highlighted the key factors, IoT and artificial intelligence, which have made the software easy-to-use in case of financial analysis. Meredith *et al.* [18] further discussed about the enhanced audit quality through decision support system and also believed that the DSS specialist and audit specialist should collaborate for effective knowledge management and decision making. Xu *et al.* [19] reviewed that management decision support systems can help in the increase of financial performance of hotels with the help of positive factors. Also, in the field of agriculture, a decision support system is proposed by Hafezalkotob *et al.* [20] to help in the decision making for olive harvesting machines. From the literature, it can be studied that the primary focus of researchers has been on the enablers and various driving forces that help in improving the implementation of DSS. The focus of various researchers has been on how technology, and its variants are the key to smart city, therefore, the main focus of this study is to focus on how this technology also delivers challenges.

Smart city has different challenges in implementing DSS so that it can meet the needs of urban population and development of city smartly [21]. These challenges include complex problems, which can be solved with the help of various approaches. Yadav *et al.* [22], in order to develop a sustainable smart city, used hybrid best–worst method (BWM) to find the influence and interpretive structural modeling (ISM) approach to develop a framework. For the evaluation of sustainability indicators [23] used fuzzy and fuzzy AHP method for the making of smart city. Another study, which was done based on the need to compare the significant efforts to make cities sustainable, was done using Analytic network process (ANP)-TOPSIS [24]. Several models, which have been prepared in order to build a smart city [25], compared two smart city planning models using MADM approach. Challenges, which were identified by experts in adopting IoT, were evaluated with MICMAC-ISM approach, which helped in developing relationships and facilized the barriers in implementing IoT [26]. Also, Rad *et al.* [27] explored the concept of u-city to overcome the urban conditions with the help of ANP and the decision-making trial and evaluation laboratory (DEMATEL) method. government and stakeholders also play an important role, which has been highlighted with the help of integrated conceptual model applied in Vienna smart city strategy [28]. Bhunia *et al.* [29] used a fuzzy approach for the implementation of e-healthcare services in smart cities. MNC also faced logistical barriers, which includes the improper transportation, and the infective flow of information, which became a problem in building of a smart city, which was then evaluated with the help of ISM-MICMAC technique [30], we can understand that various methods are used but not much research has been done using fuzzy TOPSIS. In this study, we will use the technique fuzzy-TOPSIS in order to find out the challenges in implementing DSS in smart cities as fuzzy-TOPSIS helps us to estimate various alternatives against selected standards. Based on the literature gaps as stated above, the predominant focus of this study is to develop smart cities by providing the solutions to the challenges in DSS. Despite DSS being highly efficient system, it continues to pose security threats as it is not entirely transparent. This paper highlights such challenges and delivers various alternative solutions using the technique fuzzy TOPSIS.

6.3 Research Methodology

6.3.1 Identification of Challengers and Their Solution

Challenges which come in the way of implementing DSS in smart cities identifies with the help of various literatures. This is the first step, and the challenges were identified from resources like "Google Scholar," "Science Direct," "Taylor and Francis" with the help of keywords "DSS," "smart cities," "challenges in implementing DSS," "fuzzy TOPSIS." Postextensive literature research, 11 challenges were identified and analyzed (Table 6.1).

For every challenge, a solution was proposed, and with these, five solutions were created (Table 6.2) to counter the challenges that hinder the process of using DSS in smart cities.

A panel comprising of 10 experts like assistant professor and professor from esteemed university, NGO heads and head of architectures, industry experts, were referred to establish a relation between challenges and solution and how efficient a solution is for a challenger.

These challengers, after identification, were researched upon. To overcome these challengers, a list of five solutions were identified from various journals. The list of the challengers and their description is mentioned in Table 6.2.

Table 6.1 List of challenges in implementing DSS in smart cities.

S. no.	Challenges	Description	References
1	Lack of citizen-centric approach	The cities are focusing more on technology oriented cities and hence are leaving some communities behind which result in digital divide.	[1]
2	Problem in integrating data and managing quality	Despite technological advancements, the Decision Support System (DSS), is still unable to comprehend and solve the structural inconsistencies in the data sets.	[1]
3	Privacy concern regarding data	The massive data collected from the city residents, stir up a privacy concern as the personal data shared could be compromised.	[1]

(Continued)

Table 6.1 List of challenges in implementing DSS in smart cities. (*Continued*)

S. no	Challenges	Description	References
4	Funding and financial constraints in building smart cities	The infrastructural projects are withheld due to tight budgets and unavailable investors.	OWN
5	Lack of technology knowledge in users	Users are unaware of the new technologies and pose a problem when it comes to operating the smart city services.	OWN
6	System design failure	Fundamental missteps, bugs and lack of flexibility cause the system or software to crash in unforeseeable circumstances.	OWM
7	Unaware of assumptions and unanticipated effects	There are regular updates in the system to keep the bugs away. These continuous changes make it difficult for decision makers to predict everything in advance.	OWN
8	Lack of transparency & liability	The softwares which is used in building the smart cities aren't transparent and hence could be held liable for the shortcoming in the project.	[31]
9	Connected devices and smart objects issue	Smart devices should be connected with each other in the cloud services. The embedded devices find it to be a difficult task due to network hindrances.	[32]
10	IoT challenges in storage and data	With so much information present in the cloud, it is difficult to collect and save all of it due to lack of storage.	[32]
11	Problem in having clearances in timely manner	Timely clearance of all documents and permissions for the projects doesn't happen due to the presence of bureaucrats.	OWN

Table 6.2 List of challenges in implementing DSS in smart cities.

S. no.	Solutions	Description	References
1	data assembling & usage	Big data collected should be citizen driven rather than ICT. All the data gathered should be utilized and segregated according to the departments, which should be checked using various algorithms at every stage to maintain data quality.	[33]
2	proper software usage	Availability of proper internet services with everyone having a smartphone which has strengthened passwords, firewalls and enhanced privacy makes cloud computing secure. The use of software should be done after analyzing its limitations, needs and design.	[34]
3	budget building and finding third party investors	Smart city and software require a lot of investment and hence building a budget beforehand and finding investors is a crucial step.	[35]
4	govt regulations	An effective all way coordination amongst inter-governmental bodies is necessary. This can be done via online portal which will lead in transparency and mapping.	[31]
5	company/party heading DSS department	Online procedure to have transparency for things to be done timely. Apart from this survey should be regularly updated with creation of awareness so no misuse of the software and technology is done.	[32]

6.3.2 Fuzzy TOPSIS Methodology

In this research, the evaluation of challenges faced by smart cities in implementing DSS is considered. There are various researches which has been done identifying the challenges in smart cities but not much attention is paid in implementing a method best suited to solve those challenges. We have researched on DSS as a method which can be implemented in smart cities and its challenges. Here, we are using the fuzzy-technique for order preference with similarity to ideal solution (F-TOPSIS), a multicriteria decision making (MCDM) tool, which is used to solve ranking and justification problem. There are many researchers who have used the F-TOPSIS in different areas of research for evaluation based on conflicting criteria. F-TOPSIS is used to make the criteria analysis easy for evaluation The vagueness in the scenario of real-life decision making comes into play because there are many details that are not available in exact measure. To the best of our knowledge there are no studies which have evaluated quantitatively the challenges which helped in implementing DSS in the smart cities. Moreover, the fuzzy decision-making scenario is also not included in many situations. Figure 6.1 gives the detail of the F-TOPSIS framework of evaluation.

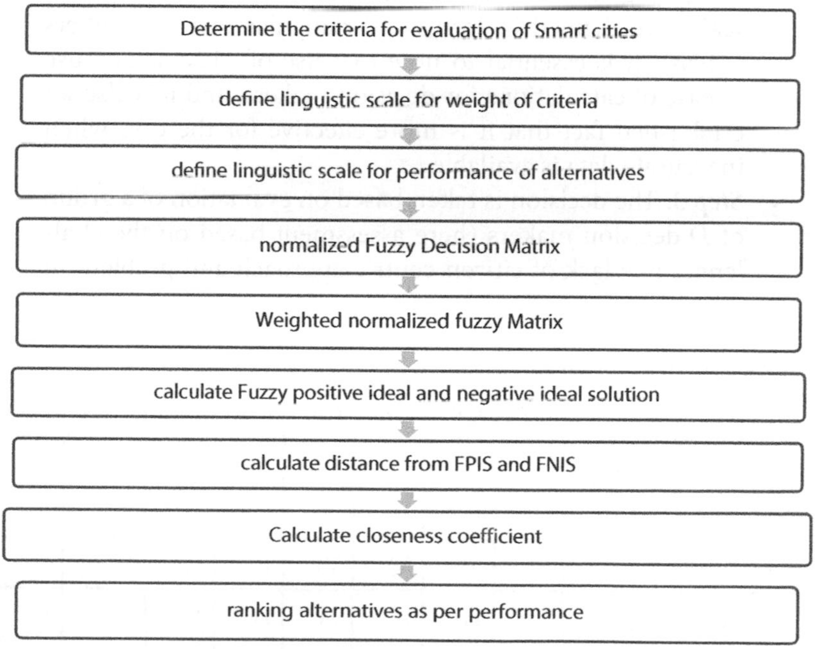

Figure 6.1 Fuzzy-TOPSIS methodology.

Steps of F-TOPSIS for ranking of challenges of smart cities

Let us say that the decision group consists of D members and the ith alternate if on jth criteria, the fuzzy rating and weight of the D decision makers about the ith alternative if on jth criteria. We describe how F-TOPSIS is used in detail for the particular case of evaluation smart cities. The implementation of DSS will help in making these smart cities more efficient, stable, and adaptable. Additionally, it is noteworthy that the current evaluation is carried out on the basis of implementation of DSS strategy.

- ***Step 1***: Initially, it is essential to determine the criteria for evaluation carefully.
  Here, the criteria are selected as discussed in the earlier section, and we have determined the (j=1,2,…,n') criteria, i.e., here we have 10 criteria (n'=10) for evaluating the (i=1,2,…,m') six alternatives (here m'=6). These criteria are determined based on the literature and in discussion with the decision makers of the firm.
- ***Step 2***: The next step is to determine the linguistic scale for evaluation of the criteria and then subsequently the alternatives. The scale used for evaluation is given in Table 6.2. Table 6.2 gives scale for evaluation of weights to criteria. The scales are defined in terms of Triangular Fuzzy Numbers (TFNs). It is essential to note that use of TFNs is because of ease of calculations for decision makers, and it is also an established fact that it is more effective for the case when inaccurate data is available.
- ***Step 3***: The decision is taken based on evaluation of a group of D decision makers (here assessment based on the challenges like lack of citizen centric approach and problem in

Table 6.3 Alternative rating.

	CR1	CR2	CR3	CR4	CR5	CR6	CR7	CR8	CR9	CR10	CR11
AL1	M	VH	L	VL	VL	L	VL	M	VL	H	M
AL2	H	H	VH	VL	L	H	H	VH	H	L	H
AL3	L	L	L	M	M	M	VL	L	H	M	L
AL4	VH	H	VL	VH	H	L	M	H	L	VL	H
AL5	M	M	H	L	L	M	H	VH	H	L	L

Table 6.4 Decision matrix for pairwise table.

	CR1	CR2	CR3	CR4	CR5	CR5	CR7	CR8	CR9	CR10	CR11
AL1	(0.26, 0.47, 0.68)	(0.46, 0.635, 0.78)	(0.22, 0.41, 0.615)	(0.185, 0.34, 0.545)	(0.11, 0.285, 0.48)	(0.1, 0.265, 0.46)	(0.15, 0.305, 0.515)	(0.35, 0.57, 0.7277777778)	(0.19, 0.37, 0.59)	(0.505, 0.74, 0.895)	(0.22, 0.39, 0.6)
AL2	(0.28, 0.48, 0.58)	(0.38, 0.61, 0.79)	(0.6, 0.81, 0.94)	(0.09, 0.24, 0.4166666667)	(0.285, 0.47, 0.665)	(0.47, 0.68, 0.85)	(0.44, 0.645, 0.805)	(0.425, 0.615, 0.775)	(0.41, 0.61, 0.805)	(0.23, 0.435, 0.625)	(0.245, 0.43, 0.615)
AL3	(0.21, 0.31, 0.5)	(0.225, 0.405, 0.6)	(0.13, 0.32, 0.53)	(0.44, 0.645, 0.805)	(0.23, 0.435, 0.625)	(0.56, 0.49, 0.71)	(0.37, 0.555, 0.73)	(0.1, 0.24, 0.455)	(0.215, 0.365, 0.565)	(0.2, 0.38, 0.61)	(0.365, 0.54, 0.73)
AL4	(0.395, 0.605, 0.78)	(0.13, 0.305, 0.505)	(0.24, 0.43, 0.62)	(0.225, 0.41, 0.59)	(0.41,0.615, 0.795)	(0.07, 0.22, 0.44)	(0.18, 0.43, 0.59)	(0.405, 0.595, 0.75)	(0.21, 0.39, 0.595)	(0.165, 0.3, 0.505)	(0.355, 0.53, 0.69)
AL5	(0.425, 0.61, 0.775)	(0.41, 0.635, 0.81)	(0.38, 0.57, 0.765)	(0.3, 0.465, 0.655)	(0.17, 0.37, 0.56)	(0.325, 0.51, 0.755555555)	(0.24, 0.455, 0.64)	(0.535, 0.75, 0.89)	(0.445, 0.675, 0.83)	(0.16, 0.35, 0.54)	(0.35, 0.54, 0.74)

data management) (D=1,2,..., D) (D=10). The group of 10 is asked to give suitable weights for the all the j criteria. The average weights for each criterion and challenges in implementing DSS w.r.t. to each criterion from the group are obtained based on the average value for 10 decision makers. Table 6.3 gives the rating of one decision makers.

- ***Step 4***: On the basis of the contextual relationship between challenges and solution, the values from the group of 10 academicians, NGO experts, etc. we obtain the final fuzzy decision matrix as well as the corresponding vector of weights. Table 6.4 gives the average rating of all the decision makers.
- ***Step 5(a)***: To ensure that the triangular fuzzy numbers lie in the range [0,1] and to compare heterogeneous criteria, the process of normalization is carried out by aggregating the fuzzy decision matrix. In the process of normalization, diverse comparisons have to be used to benefit criteria and two cost criteria. Table 6.5 shows the normalized matrix.
  (b) The next step requires the fuzzy normalize decision matrix to be weighted. Each criterion has a weight. These weights are then multiplied by the fuzzy normalized decision matrix, values obtained for the smart cities. These calculations are done using fuzzy multiplication rule. The matrix obtained is the weighted normalized fuzzy matrix. Table 6.4 shows the weighted normalized fuzzy decision matrix.
- ***Step 6***: in this step, the fuzzy positive ideal solution (FPIS) and the fuzzy negative ideal solution (FNIS) values are computed

$$\text{FPIS} = A^{*} = (1,1,1)$$

Table 6.5 Relative-closeness matrix.

	CCj1	CCj2	CCj3	CC	Rank
AL1	0.3466729019	0.5329034127	0.5701596408	20.0673991	3
AL2	0.4489162988	0.4527088791	0.5485102795	18.67534032	4
AL3	0.4722127411	0.576386451	0.6323732255	22.587874	1
AL4	0.426337983	0.576960149	0.6343500001	22.24031841	2
AL5	0.3504180762	0.4057283371	0.51156909	16.42906818	5

$$\text{FNIS} = A^{-} = (0,0,0)$$

- ***Step 7***: in this step, the distance computed from each alternative to the FPIS and to the FNIS. d_i^* and d_i^- are the distances calculated from the FPIS and FNIS, respectively, for each of the manufacturing plants.
- ***Step 8(a)***: In the final step, the closeness coefficient (CC_i) for each alternative is calculated.
  (b): The closer the value of CC_i is to unity, the better the alternative. Arranging (CC_i) values in descending order, we obtain the top ranked alternatives [36].

The F-TOPSIS method is a MCDM method, which has been used in this research in order to preferentially rank the solutions, which help in implementing DSS in smart cities. The fuzzy nature helps to incorporate the vagueness in decision making in the real life. The idea of TOPSIS is to achieve the ranking of the alternatives by evaluating their performance with respect to the criteria (here the use of DSS criteria). The closeness coefficient value for a particular alternative signifies how well it is performed as close to the ideal alternative (here each challenge with the best solution). The alternative with the highest value of closeness coefficient is the best alternative for implementing DSS strategy in the smart city. The alternative with the lowest value of closeness coefficient is the worst performing alternative.

6.4 Case Background

A DSS is an automated program used to help conclusions, decisions, and strategies in an association or a business. A DSS filters through and dissects monstrous measures of information, accumulating far-reaching data that can be utilized to tackle issues and in dynamic.

Their components provide a strategic and tactical approach to deal as it has the required tools to plan and execute the necessary part of the projects since DSS have an ability to deal with big data, which is there in the case of smart city.

Fokker *et al.* [37] researched that it is very important to forecast and plan a smart city architecture with a long-term vision. So, for that, DSS is quite helpful as it provides insights into the future and has the ability to deal with big data.

But according to Filip [38], DSS poses a lot of challenges in its implementation. It is also seen in the research that the DSS has several challenges (Table 6.1) and to overcome these challenges, a list of five solutions is proposed. To establish the relationship and its strength, a panel of 11 experts, comprising academicians, NGO heads, and architects, were asked.

Furthermore, to find the ranking of these five solutions, and find out which is the most optimized method, the approach of fuzzy TOPSIS was used.

This bring us to the following objectives:

- To find a set of solutions for the challenges in implementing DSS in smart cities.
- To find the most optimal solution to these challenges so that smart cities work in the favor of their residents.

6.5 Case Description

Fuzzy TOPSIS can be used to evaluate several options based on selected criteria. In the TOPSIS approach, an option is chosen as the optimal one that is closest to the fuzzy-positive ideal solution (FPIS) and fuzzy-negative ideal solution (FNIS). Best performance reviews are shown by FPIS, whereas worst performance reviews are shown by FNIS.

The steps adapted to present fuzzy TOPSIS are as follows.

- In the first step of fuzzy TOPSIS, it consists of the pooling the performance judgments and weights given by the decision makers. In our study, we have taken a group of 11 decision makers, which consist of academicians, experts from industries, startup industries, NGO, etc. Here, we have 10 challenges with five alternative solutions. Fuzzy multicriteria decision matrix X can be shown in Table 6.3.

In the above table, the decision matrix of one decision maker is shown. Similarly, the other decision maker reviews are recorded as shown in Table 6.4.

- Apply fuzzy numbers — conversion scales are applied as it is difficult what decision makers designate a precise value for the criteria, therefore, they use the linguistic terms into fuzzy numbers.

In step 2—the aggregated fuzzy ratings for alternatives and aggregated fuzzy ways for criteria are computed.

Step 3—in this step, the fuzzy decision matrix is then converted into normalized fuzzy decision matrix, so that the numbers obtained in different scales are comparable.

Step 4—the weighted normalized fuzzy matrix is obtained by multiplying the normalized fuzzy matrix with the associated weight criteria corresponding to that column.

Step 5—The next step is to determine the fuzzy positive ideal solution given by S+ and fuzzy negative ideal solution given by S– from the weighted normalized fuzzy decision matrix. After the FNIS and FPIS are calculated, the distance from each alternative to FNIS and FPIS are computed which is shown in the table as A+ and A–.

After finding the distance, the relative closeness to the ideal solution for each alternative is calculated. The relative closeness of jth alternative w.r.t A+ is shown in Table 6.5 above.

The index value of Cj lies between 0 and 1. The alternative that has the highest coefficient closeness is the best solution, whereas the lowest coefficient indicate the worst solution.

In the similar way the ranking of the highest to lowest solutions are done which can be shown in Table 6.5.

6.6 Result Discussion

From the CC table (Table 6.5), the results for the most optimal solution are obtained from the rank. The solution having rank 1 is considered the most suitable and the solution having rank 5, the least suitable to overcome the challenges.

The solution having rank 1 is "budget building and finding third party investors." According to analysis done using the method of fuzzy TOPSIS, the solution having rank 1 is considered the best to deal with the challenges. Hence, budget building and finding third-party investors, which is associated with proper planning of funds and finding third-party investors to finance the whole process of building smart cities using DSS is the best way to overcome the barriers.

The second-best solution, having rank 2, is "government regulations." Government regulations are concerned with dealing with bureaucracy, transparency, delay in getting clearances. These problems are some of the biggest problems, which hamper the use of DSS in building the smart cities.

The third best solution is "data assembling and usage." There is an involvement of a big data when structuring smart cities. There is a lot of confidential information in the data, which if leaked could hamper the personal lives of the citizens. Apart from that, if the software is not used properly then the results from the data will not be accurate, and the end result of the city could be wrong. Hence, it is important to deal with the data carefully and use the software accurately.

The solution on fourth is "proper software usage," which indicates the mishandling of software and provides the key for the same so that the end result is not hampered.

Least important solution, which has rank 5 is, "company/party heading DSS department." The company/third party that is handling DSS and managing the software needs to pay attention not only to the software but also for whom it is used, it is catering to their needs and the outcome from the software should be appropriate for use.

6.7 Conclusion

Smart cities are the future of India. They aim to be sustainable, contribute to economic and infrastructural development and maximize the use of information communication technology. They create a pool of opportunities for architect and workers and improve the standard of living and also improve other quality factor of the residents. To develop smart cities, data from residents, builders, and other concerned people need to be collected. Hence, to build smart cities, a lot of big data are involved.

Analyzing big data requires the help of software and experts. The use of DSS to analyze this big data and to help build a smart city is discussed in this paper. There are sever challenges which come in the way, when it is related to using DSS in smart cities. This paper aims to answer two questions mentioned: it is to find the challenges and formulate a group of solutions to these challenges and furthermore to evaluate using fuzzy TOPSIS as to which solution is the best optimized and the which is the least optimized.

This is why this research has been conducted. Eleven challengers were identified (Table 6.1) via literature and their solutions (Table 6.2). To establish a relationship between the challengers and solutions, a panel of experts were questioned, which included academicians, industry experts, NGO heads, scholars. The technique of fuzzy TOPSIS has been implemented to obtain the results. The following results & observation were attained:

As per the CC table (Table 6.5), the solution which holds rank 1 is considered the most suitable solution to over the challenges discussed in the paper and the solution which holds rank 5 is considered the least suitable solution to deal with challenges.

The solution which holds rank 1 is "budget building and finding third party investors," which is related to efficient fund planning and finding third-party investor to finance the whole project of smart cities. Solution government regulations' holds rank 2, which is concerned with eradicating governmental procedures and bureaucracy which delay the whole process of smart city building. Solution with rank 3 is "data assembling and usage." This is concerned with proper handling of data as a lot of confidential information might be leaked if not handled carefully, and using it appropriately to get the best results. Solution on position 4 is "proper software usage," which talk about using the software appropriately and having full knowledge before it is handled. The last one is "company/party heading DSS department." This deals with the third party who will be operating the software. They need to be careful of the needs of the citizens and accordingly the software should be used.

Our study about challenges and their solution in using DSS in smart cities is for the northern part of the country, India. Manager, academicians, builders and architects can change the result depending on the dimensions and environment.

References

1. Lim, C., Kim, K.J., Maglio, P.P., Smart cities with big data: Reference models, challenges, and considerations. *Cities*, 82, 86–99, 2018.
2. Xie, J., Tang, H., Huang, T., Yu, F.R., Xie, R., Liu, J., Liu, Y., A survey of blockchain technology applied to smart cities: Research issues and challenges. *IEEE Commun. Surv. Tut.*, 21, 3, 2794–2830, 2019.
3. Ismagilova, E., Hughes, L., Dwivedi, Y.K., Raman, K.R., Smart cities: Advances in research—An information systems perspective. *Int. J. Inf. Manage.*, 47, 88–100, 2019.
4. Lea, R.J., Smart cities: An overview of the technology trends driving smart cities, IEEE, 2017.
5. Chichernea, V., The use of decision support systems (DSS) in smart city planning and management. *JISOM*, 8, 2, 1–14, 2014.

6. Biswas, K. and Muthukkumarasamy, V., Securing smart cities using blockchain technology, in: *2016 IEEE 18th International Conference on High Performance Computing and Communications; IEEE 14th International Conference on Smart City; IEEE 2nd International Conference on Data Dcience and Dystems (HPCC/SmartCity/DSS)*, IEEE, pp. 1392–1393, December 2016.
7. Yaqoob, I., Hashem, I.A.T., Mehmood, Y., Gani, A., Mokhtar, S., Guizani, S., Enabling communication technologies for smart cities. *IEEE Commun. Mag.*, 55, 1, 112–120, 2017.
8. Anagnostopoulos, T., Zaslavsky, A., Kolomvatsos, K., Medvedev, A., Amirian, P., Morley, J., Hadjieftymiades, S., Challenges and opportunities of waste management in IoT-enabled smart cities: A survey. *IEEE Trans. Sustain. Comput.*, 2, 3, 275–289, 2017.
9. Wenge, R., Zhang, X., Dave, C., Chao, L., Hao, S., Smart city architecture: A technology guide for implementation and design challenges. *China Commun.*, 11, 3, 56–69, 2014.
10. Luckey, D., Fritz, H., Legatiuk, D., Dragos, K., Smarsly, K., Artificial intelligence techniques for smart city applications, in: *International Conference on Computing in Civil and Building Engineering*, Springer, Cham, pp. 3–15, August 2020.
11. Anthopoulos, L.G., The smart city in practice, in: *Understanding Smart Cities: A Tool for Smart Government or an Industrial Trick?*, pp. 47–185, Springer, Cham, 2017.
12. Park, E., Del Pobil, A.P., Kwon, S.J., The role of Internet of Things (IoT) in smart cities: Technology roadmap-oriented approaches. *Sustainability*, 10, 5, 1388, 2018.
13. Bawa, M., Caganova, D., Szilva, I., Spirkova, D., Importance of internet of things and big data in building smart city and what would be its challenges, in: *Smart City 360°*, pp. 605–616, Springer, Cham, 2016.
14. Medvedev, A., Fedchenkov, P., Zaslavsky, A., Anagnostopoulos, T., Khoruzhnikov, S., Waste management as an IoT-enabled service in smart cities, in: *Internet of Things, Smart Spaces, and Next Generation Networks and Systems*, pp. 104–115, Springer, Cham, 2015.
15. Kashevnik, A. and Lashkov, I., Decision support system for drivers & passengers: Smartphone-based reference model and evaluation. *2018 23rd Conference of Open Innovations Association (FRUCT)*, 2018.
16. Papastamatiou, I., Marinakis, V., Doukas, H., Psarras, J., A decision support framework for smart cities energy assessment and optimization. *Energy Proc.*, 111, 800–809, 2017.
17. Chatterjee, S., Kar, A.K., Gupta, M.P., Success of IoT in smart cities of India: An empirical analysis. *Gov. Inf. Q.*, 35, 3, 349–361, 2018.
18. Meredith, K., Blake, J., Baxter, P., Kerr, D., Drivers of and barriers to decision support technology use by financial report auditors. *Decis. Support Syst.*, 139, 113402, 2020.

19. Xu, X., Zhang, L., Baker, T., Harrington, R.J., Marlowe, B., Drivers of degree of sophistication in hotel revenue management decision support systems. *Int. J. Hosp. Manage.*, 79, 123–139, 2019.
20. Hafezalkotob, A., Hami-Dindar, A., Rabie, N., Hafezalkotob, A., A decision support system for agricultural machines and equipment selection: A case study on olive harvester machines. *Comput. Electron. Agr.*, 148, 207–216, 2018.
21. Rathore, M.M., Ahmad, A., Paul, A., IoT-based smart city development using big data analytical approach, in: *2016 IEEE international conference on automatica (ICA-ACCA)*, IEEE, pp. 1–8, October 2016.
22. Yadav, G., Mangla, S.K., Luthra, S., Rai, D.P., Developing a sustainable smart city framework for developing economies: An Indian context. *Sustain. Cities Soc.*, 47, 101462, 2019.
23. Anand, A., Rufuss, D.D.W., Rajkumar, V., Suganthi, L., Evaluation of sustainability indicators in smart cities for India using MCDM approach. *Energy Proc.*, 141, 211–215, 2017.
24. Ozkaya, G. and Erdin, C., Evaluation of smart and sustainable cities through a hybrid MCDM approach based on ANP and TOPSIS technique. *Heliyon*, 6, 10, e05052, 2020.
25. Mattoni, B., Pompei, L., Losilla, J.C., Bisegna, F., Planning smart cities: Comparison of two quantitative multicriteria methods applied to real case studies. *Sustain. Cities Soc.*, 60, 102249, 2020.
26. Janssen, M., Luthra, S., Mangla, S., Rana, N.P., Dwivedi, Y.K., Challenges for adopting and implementing IoT in smart cities: An integrated MICMAC-ISM approach. *Internet Res.*, 29, 6, 1589–1616, 2019.
27. Rad, T.G., Sadeghi-Niaraki, A., Abbasi, A., Choi, S.M., A methodological framework for assessment of ubiquitous cities using ANP and DEMATEL methods. *Sustain. Cities Soc.*, 37, 608–618, 2018.
28. Fernandez-Anez, V., Fernández-Güell, J.M., Giffinger, R., Smart City implementation and discourses: An integrated conceptual model. The case of Vienna. *Cities*, 78, 4–16, 2018.
29. Bhunia, S.S., Dhar, S.K., Mukherjee, N., iHealth: A fuzzy approach for provisioning intelligent health-care system in smart city, in: *2014 IEEE 10th International Conference on Wireless and Mobile Computing, Networking and Communications (WIMOB)*, IEEE, pp. 187–193, October 2014.
30. Gandhi, N., Haleem, A., Shuaib, M., Kumar, D., Analysis of logistical barriers faced by MNCs for business in Indian smart cities using ISM-MICMAC approach, in: *Smart Cities—Opportunities and Challenges*, Springer, Singapore, 2020.
31. Rana, N.P., Luthra, S., Mangla, S.K., Islam, R., Roderick, S., Dwivedi, Y.K., Barriers to the development of smart cities in Indian context. *Inf. Syst. Front.*, 21, 3, 503–525, 2019.
32. Evan, C., Internet of things challenges in storage and data, 2018. https://www.computerweekly.com/news/252450705/Internet-of-things-challenges-in-storage-and-data?amp=1 [Accessed on 04 May 2021].

33. Zhuang, C., Gong, J., Liu, J., Digital twin-based assembly data management and process traceability for complex products. *J. Manuf. Syst.*, 58, 118–131, 2021.
34. Kumar, S., An assessment of research process management software usage in Indian higher education, in: *Decision Analytics Applications in Industry*, pp. 409–419, Springer, Singapore, 2020.
35. Verdecchia, R., Kruchten, P., Lago, P., Malavolta, I., Building and evaluating a theory of architectural technical debt in software-intensive systems. *J. Syst. Software*, 176, 110925, 2021.
36. Kore, M.N.B., Ravi, K., Patil, A.P.M.S., A simplified description of fuzzy TOPSIS method for multi criteria decision making. *Int. Res. J. Eng. Technol.*, 4, 5, 2047–2050, 2017.
37. Fokker, E.S., Koch, T., Dugundji, E.R., *Long-Term Forecasting of Off-Street Parking Occupancy for Smart Cities (No. TRBAM-21-01724)*, 2021.
38. Filip, F.G., DSS—A class of evolving information systems, in: *Data Science: New Issues, Challenges and Applications*, pp. 253–277, Springer, Cham, 2020.

7

Evaluation of Criteria for Implementation of Capabilities in a Smart City's Service Supply Chain: A Teacher Trainer's Perspective

Vasundhara Kaul[1] and Arshia Kaul[2]*

[1]Carpediem EdPsych Consultancy LLP, Mumbai, India
[2]ASMSOC, NMIMS University, Mumbai, India

Abstract

Change is the only constant, they say; with technology, the changes seem to be happening ever so often. The question that arises is to what extent have the changes been implemented and what are the capabilities of systems in various sectors to implement these changes. COVID-19 is a medical catastrophe, but it has come with a silver lining. It has forced everyone to adapt to using technology. People in different sectors were forced to ensure that they learnt the new ways of working. Work from home became the norm, which in turn led to the imperative nature of connectivity through the online platforms. Smart cities which were only in the planning and policy development stage suddenly saw a "jugaad" technology playing a role in it. This "jugaad" technology will work for the interim period and will most definitely initiate the start of complete connectedness between individuals and companies and in turn developing smart cities. In this chapter, the aim is to focus on understanding the barriers of use of technology in the education sector. There are many stakeholders which can be considered in the education sector. It is impossible to consider all dimensions in one study. Our aim is to determine the barriers faced by teacher trainers while training in the online mode. We use a multicriteria decision making tool DEMATEL to further establish the cause-and-effect group between the criteria. It is necessary to remove the antecedents of the consequents and establish a better education system and lead to smart education for smart cities.

**Corresponding author*: arshia.kaul@gmail.com

Loveleen Gaur, Vernika Agarwal and Prasenjit Chatterjee (eds.) Decision Support Systems for Smart City Applications, (119–136) © 2023 Scrivener Publishing LLC

Keywords: Smart education, smart cities, antecedents, consequents, DEMATEL

7.1 Introduction

Supply chain management (SCM) has become very important for business organizations to remain competitive in the global marketplace. There has been extensive research, which has been done in this direction by many researchers from time immemorial [1, 2]; [3] till date [4]. There were many who focused on the product flow of the supply chain [5], and supply chain was traditionally categorized as the branch of operation management [5–8]. It initially focused on manufacturing and not on the services. In recent times, the service industry has become very important in the global economy. This has evidence from the many manufacturing organizations such as GM and IBM which have gained more from their service units [8]. There is much value added from the service division of the organizations [9]. There was limited research initially which concentrated on the service sector. The distinction between the service supply chains (SSC) and manufacturing supply chains is based on structure and managerial aspects. There are also different roles that the customer plays in an SSC. The foundation of the SSC is on the Unified Service Theory (UST), which defines that the distinguishing factor of services is that they possess a bidirectional supply chain relationship. In such a situation the customers also act as suppliers [10, 11]. In any SSC, the customers also supply some input materials, labor, specifications as a supplier. There is an expansion in the role of the customer in an SSC.

One service area that recently gained importance is the education sector. SCM in the academia also called the educational supply chain is for ensuring that good quality education may be imparted to the students and this in turn would help in the development of the society. The current research concentrates on the education sector being considered as a service sector. We draw an analogy between any service supply chain and the education sector organization. Figure 7.1 highlights the analogy between a general service supply chain and education sector [12].

In the education sector, there are various stakeholders, and these stakeholders can be classified as those providing service of education and those receiving the service of education. A research by Ellram *et al.* [13], identified the seven processes, such as the information flow, capacity and skills management, demand management, supplier relationship management, and cash flow from the traditional supply chain. This study was extended by Baltacioglu *et al.* [14] to the seven processes for service supply chain, such as demand, capacity and resources, customer relationship, supplier relationship,

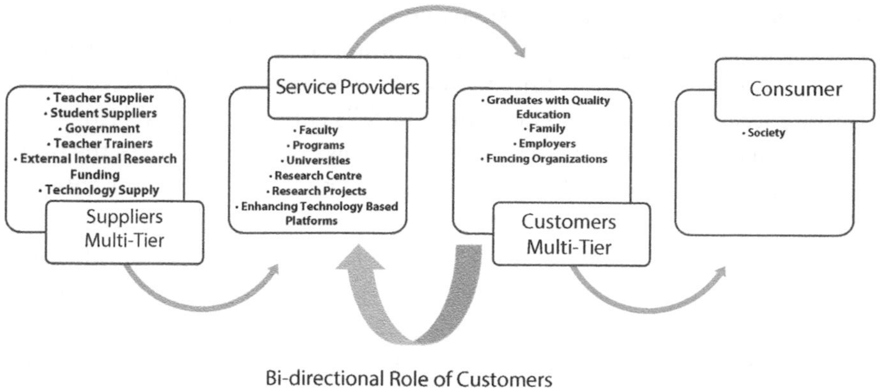

Figure 7.1 Educational supply chain. Source: modified by authors based on version from [12].

order process, service performance, and information and technology management. Borrowing from Baltacioglu *et al.* [14] and the discussion on processes of service organization, in our paper, we concentrate on processes of an educational organization, such as the capacity and resources management and the information and technology management more closely. While we consider the educational supply chain in context of information and technology management, we tend toward developing the processes for development of a smart connected city. Through the connected educational supply chain, we wish to propose changes in the educational systems like never seen before. With these changes implemented, the educational supply chain would lead to develop workforce eligible to work in a well-connected smart city.

With more private players coming into the market for education, there is also a lot of competition in this sector. All the customers in such a situation are more aware, and since they have many options, the moment they do not get value from the organization which they are enrolled in, they have many more options to move to. These students also move or otherwise are very demanding of the system which they belong to. Therefore, the educational institutions need to provide exclusive services to the students (customers of the given system). For providing the best education for their students in the connected world, the educational institutions need to ensure that their faculty members (major stakeholders) are well equipped. Training in this direction need to be given in to the faculty members. This is a continuous process and although the pandemic got in road blocks to the functioning, those who quickly adapted have coped as best possible in these situations. Although the connectedness due to pandemic was brought in under dire circumstances, in the long term, we will be able to help to lead to development of smart connected educational systems. The trainings also needed to be conducted online in the changed situations. Therefore, it is essential to understand what were the pitfalls faced while training teachers online and how these pitfalls could be tackled [15].

The rest of the chapter is structured as follows: section 7.2 gives the literature review in the context of educational supply chain and highlights the research gap and motivation. Section 7.3 defines the objectives. Section 7.4 defines the problem definition. In section 7.5, we present the evaluation of the real-life case with the help of DEMATEL based on the opinions of the decision makers. Section 7.6 concludes the chapter and defines a few limitations of the research, which will lead to the future scope of the study.

7.2 Literature Review

In this section, we highlight the literature in the context of our research. We present the literature in three distinct sections: (i) service supply chain, (ii) educational supply chain, (iii) evaluation of educational supply chain.

7.2.1 Service Supply Chain

There is some research which has concentrated on the service supply chain. In the extant literature, Bolliger and Wasilik [15] and Nie and Kellogg [16] discussed the breadth of the customer involvement in the SSC. The concept of bidirectional supply chain is discussed as part of UST by Sampson [10]. The concept of bidirectional supply chain is that the customers provide resources before the service providers could provide what the customer needs. The service providers have the capability of providing services independently, but they need resources from the customers [11]. The resources provided by the customers as per the UST is the customer's self (mind, body, and effort), some of their belongings or information [10]. The customer-as-resource-supplier distinguishes the SSC from the manufacturing supply chain as per UST [11]. Normann [17], in his study, discussed the insurance sector. In the study, the insurance client provides the personal details and details of the car based on which the insurance company determines the insurance to be provided based on the risk incurred. Lovelock [18], in their paper, categorized the services into four categories viz. services acting on customers' minds for example education, services acting on customers' bodies for example healthcare, services acting on the physical possessions of customers, such as television repairs, and the fourth, which acts on the information provided by the customers, such as the tax accounting. Customers can also assist the production of service as labor as discussed by Grönroos [19]. From the above extant literature, it can be seen that there is limited literature that has discussed education as a service. Though there have been some very recent studies, which have now started looking at education as an important service supply chain [20], [21]. The next section presents in detail how the education industry has been considered as part of the service supply chain.

7.2.2 Education as a Service Supply Chain

There has been an extensive research that has been carried out in education as a service. There are different ways of imparting knowledge and many aspects which have been considered in the literature. Moore and Kearsley [22] have discussed distance learning as the environment in case the students and teachers are separated by distance and time. In their study, Allen *et al.* [23] discussed the benefits for offering online courses. The online courses had improved student access, higher degree of completion rate, and appealing nature of online courses for nontraditional students. Rovai *et al.* [24] stated that when any structured learning is separated by time or geography, then the setup could be termed as distance learning setup. As per Hogan and McKnight [21], online learning is the one in which all the content material and interactions happen via the Internet. Further, they also discussed how important commitment of faculty is, in these situations in order to provide successful education at a distance. There are also aspects of burnout of the faculty due to online teaching as discussed by Hogan and McKnight [21].

7.2.3 Evaluation of Educational Supply Chain

For the evaluation of the educational supply chain, there have been many different tools and techniques, which have been used. Some of them we highlight briefly in this section. There have been many debates with reference to the definition of quality as highlighted by Matorera [25]. It is believed that no singular definition of quality can be taken to be complete. There are many other researchers who have linked the professional development of teachers to be impacting the teacher quality [26, 27]. Quality in educational systems is being achieved by teaching and learning as discussed by Hogan and McKnight [21]. Quality Function Deployment (QFD) is used to achieve better quality of the system for providing products and services. Further, Sagnak *et al.* [28] stated that educational institutions are considering at looking at ensuring better and better quality of the educational systems. We have highlighted only a few studies here which have considered the evaluation of the educational supply chain and the role of its stakeholders. We can observe that there is no such study to the best of our knowledge which have considered evaluation of the educational supply chain from the perspective of the teacher trainer under the connected smart city context.

7.2.4 Research Gap and Motivation

From the extant literature, we can observe that although there are some researchers who have considered the concept of educational supply chain as a specific case of the service supply chain. There are certain details of the education supply chain that have been mentioned viz., the method of teaching, the evaluation of student satisfaction, role of the educational institution, and the various stakeholders in the educational supply chain. To the best of our knowledge, there is no research which has considered (i) the perspective of teacher trainer and the problems faced by them especially while they conduct training, (ii) the current studies even if they discuss the teacher trainer perspective, they consider the discussion more conceptually, (iii) the integration of technology as an enhancer of the educational service supply chain has also not been considered while quantitative evaluation. Keeping in mind the literature gap that we have been able to identify, the following are the objectives of the study.

7.3 Objectives

1. To identify the criteria for the trainer conducting online training that are influencing or being influenced in an educational supply chain for smart city development.
2. To quantitatively segregate the criteria into cause-and-effect group from the perspective of the trainer in an educational supply chain for smart city development.
3. To identify the most important causes of the problems while conducting training online by teacher trainers and accordingly draw a plan for implementation for training in the future which will enhance the service provided by the education supply chain for smart city development.

7.4 Problem Definition

Education can be considered as a major part of the service industry. It has different characteristics from the manufacturing industry, and the outputs are more intangible as opposed to tangible. The aim of this study is to enhance the overall service output provided from the educational service supply chain. The particular case under consideration is to consider the trainer perspective and their contribution in the educational supply chain. In an educational supply chain, if considered from a teacher trainer perspective, teachers will require training, which is provided by the trainers. These teachers will equip themselves better in terms of better knowledge in their specific field based on the training (which is the process considered) and which will give the output as the better trained teachers. In the broader context, if the teachers are better trained then in turn, they can train their students better who are eventually going to become the part working force. These relationships between the stakeholders are part of a network. The educational supply chain network can be understood with the help of Figure 7.2.

The figure has two sets of the context in which we are trying to carry out this study. In the figure, the role of a teacher trainer and their contribution in enhancing the educational supply chain is highlighted. With this context in mind, it is essential to study the criteria which the teacher trainers need to consider for developing the teacher training especially in the online scenario. One thing has become very evident ever since the pandemic has set in all over the world, that things cannot stop to function. It holds true for teacher training context as well. With passing time, it has been observed that teacher training cannot be stalled till the pandemic is over. Teachers need to enhance their knowledge and equip themselves with new knowledge,

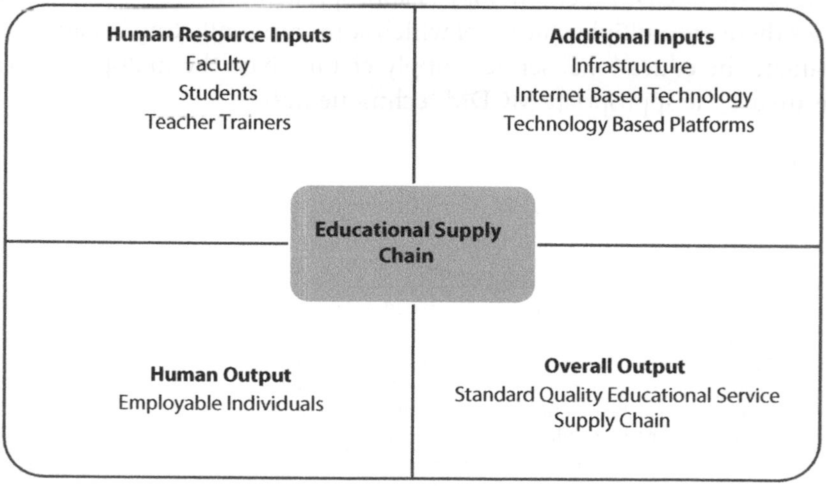

Figure 7.2 Input-output relationships in educational service supply chain. Source: modified by authors based on version from reference [12].

more so in the online teaching learning scenario. The teachers will get their knowledge from training programs, and thus, they would need trainers who are well equipped. In this scenario, the idea is to understand what the trainers need to keep in mind while developing these online training, in order to enhance the overall output of the educational supply chain.

7.5 Numerical Illustration

Given the context of enhancing the quality of the overall educational supply chain by developing good training programs for the teachers, we have taken a group of 12 experienced teacher trainers for the further case discussion of the study. These 12 experienced trainers have at least 10 years of experience each and are currently teaching in various reputed institutes of India (names not mentioned due to reasons of confidentiality). First, the criteria for developing a good training program for the future are identified in discussion with the experts. Subsequently, the criteria are divided into the two groups the cause-and-effect groups in order to consider the causation of the problems if any that exist while conducting the online training. For dividing the criteria into cause-and-effect groups, the multi-criteria decision making (MCDM) Decision making trial and evaluation laboratory (**DEMATEL**) tool is used. The details of steps are given in the following methodology section. We have only highlighted the main steps of the methodology here and the readers are encouraged to further refer to [29]. The MCDM methodology is appropriate in this real-life situation since there are multiple criteria, of which some are conflicting in nature. To evaluate the educational service supply chain, given the multiple criteria we are use the appropriate MCDM technique [29].

Methodology

Highlighting the steps of DEMATEL in Figure 7.3.

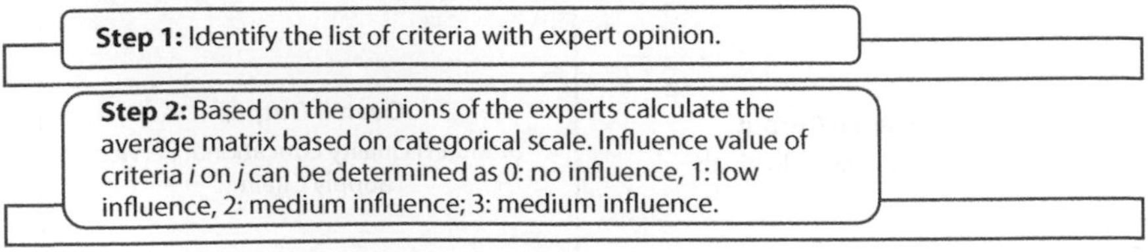
Step 1: Identify the list of criteria with expert opinion.
Step 2: Based on the opinions of the experts calculate the average matrix based on categorical scale. Influence value of criteria *i* on *j* can be determined as 0: no influence, 1: low influence, 2: medium influence; 3: medium influence.

Figure 7.3 Steps of DEMATEL. *(Continued)*

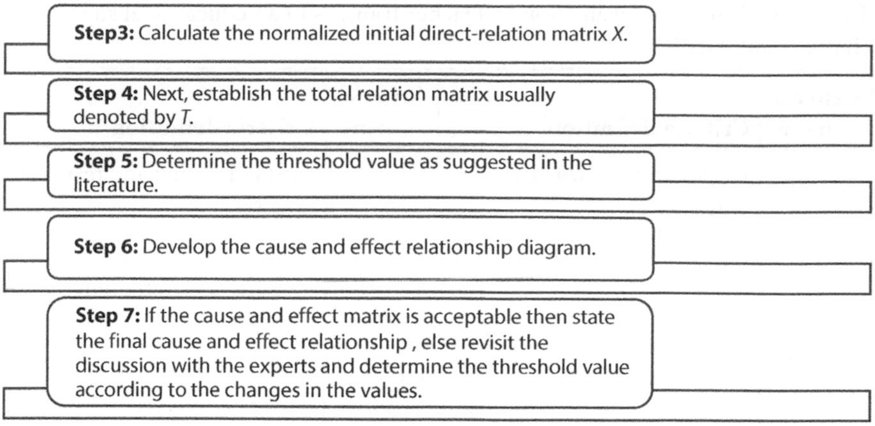
Step3: Calculate the normalized initial direct-relation matrix X.
Step 4: Next, establish the total relation matrix usually denoted by T.
Step 5: Determine the threshold value as suggested in the literature.
Step 6: Develop the cause and effect relationship diagram.
Step 7: If the cause and effect matrix is acceptable then state the final cause and effect relationship , else revisit the discussion with the experts and determine the threshold value according to the changes in the values.

Figure 7.3 (Continued) Steps of DEMATEL.

Results and Discussion

As we have considered the 12 decision makers. In this section, we show the detailed matrices of DEMATEL. Initially, we state the criteria for consideration for a teacher trainer in the context of the educational service supply chain. Table 7.1 gives the criteria, which will be evaluated by the decision makers in order to determine the final cause and effect groups. These are based on discussion with decision makers and literature [28–32].

Table 7.1 Criteria for evaluation by teacher trainer's in an educational supply chain.

Criteria name	Criteria definition	Criteria name	Criteria definition
C1	Acceptance of Participant teachers	C11	Technical difficulties other than internet issues
C2	Age of participant teachers acting as a barrier	C12	Diversity in participants
C3	Internet Connectivity/ Technology being used is reliable	C13	Level of interactions with participants
C4	Health Issues of participant teachers	C14	Flexibility in the online structure for training

(Continued)

Table 7.1 Criteria for evaluation by teacher trainer's in an educational supply chain. (*Continued*)

Criteria name	Criteria definition	Criteria name	Criteria definition
C5	Health Issues of teacher trainer	C15	Cost of location and related cost
C6	Effectiveness of Teaching	C16	Higher workload as compared to training offline
C7	Clarity in concepts	**C17**	Self-paced Online trainings can be accessed at any time (no seriousness of participants)
C8	Content sharing and access limited	C18	Creative resources used
C9	Online discipline (Cross-questioning without listening, Multiple people asking same questions repeatedly)/control on participant	C19	Better feedback can be provided
C10	Hardware Costs of Training online	C20	Compensation
		C21	Gratifying training online as there are many participants who can be reached out to

Based on the criteria as given in Table 7.1, we carry out the analysis using DEMATEL. The criteria in Table 7.1 are the input for DEMATEL and evaluation is carried out based on the same. Table 7.2 gives the initial average matrix for the 21 criteria based on the opinions of 12 decision makers.

Given the limitation of the pages, we have not included the detailed calculation tables of DEMATEL in this chapter. We have followed the steps of DEMATEL as described in the methodology section and as highlighted by Shieh *et al.* [29].

Table 7.2 Initial matrix for evaluation by one decision maker for DEMATEL.

Criteria for online training (i)--(j)	C1	C2	C3	C4	C5	C6	C7	C8	C9	C10	C11	C12	C13	C14	C15	C16	C17	C18	C19	C20	C21
C1	0	0	0	0	3	3	3	2	3	0	2	0	1	0	0	2	3	0	3	0	1
C2	3	0	0	3	3	3	3	3	3	0	3	2	3	0	0	3	2	3	0	0	2
C3	3	3	0	3	2	2	2	3	1	3	0	0	3	2	3	3	2	2	3	3	3
C4	3	3	3	0	0	3	2	0	2	0	1	2	2	0	0	2	2	0	0	0	0
C5	3	0	0	0	0	3	3	2	2	2	1	0	3	2	3	3	0	3	3	0	0
C6	3	0	0	1	0	0	3	1	3	0	1	2	3	0	0	0	0	2	2	3	0
C7	0	0	0	0	0	0	0	3	0	0	0	3	2	2	0	3	3	3	3	0	3
C8	1	1	0	0	0	1	2	0	2	0	2	0	1	1	0	3	3	2	0	0	1
C9	3	3	0	1	2	3	3	1	0	0	2	0	3	2	2	3	1	0	3	0	2
C10	0	0	2	0	0	2	2	1	1	0	1	0	1	2	2	1	2	1	0	2	0
C11	3	3	2	1	2	3	3	3	3	2	0	0	3	3	2	3	2	2	2	2	2
C12	0	0	0	1	1	3	3	2	3	3	3	0	3	3	3	3	3	3	2	1	3
C13	3	0	0	1	2	3	3	3	2	0	3	1	0	2	0	3	1	3	3	0	3

(Continued)

Table 7.2 Initial matrix for evaluation by one decision maker for DEMATEL. *(Continued)*

Criteria for online training (i)--(j)	C1	C2	C3	C4	C5	C6	C7	C8	C9	C10	C11	C12	C13	C14	C15	C16	C17	C18	C19	C20	C21
C14	2	0	0	1	1	2	2	2	2	0	1	3	2	0	2	2	2	2	2	2	2
C15	0	0	2	0	0	1	0	1	0	2	2	0	0	0	0	0	0	1	0	3	0
C16	2	1	1	0	3	2	2	1	1	0	2	2	1	2	0	0	2	2	2	2	3
C17	3	3	0	2	2	2	2	3	3	2	0	3	0	3	2	2	0	2	2	2	3
C18	3	2	2	0	0	3	3	3	3	2	2	3	3	3	3	3	2	0	3	3	3
C19	2	2	0	0	0	3	3	3	3	0	3	0	3	2	0	3	0	0	0	2	3
C20	0	0	3	0	0	0	0	2	2	3	2	0	0	0	3	1	2	3	1	0	2
C21	0	0	0	0	0	2	2	1	0	0	0	0	2	1	2	0	0	2	2	1	0

In Table 7.3, the final relationship between the criteria is taken. The relationship is defined in terms of prominence vector (R_i+C_j) and the relation vector (R_i-C_j). R_i-C_j implies that it is a receiver (effect) and R_i-C_j positive implies that it is cause.

Table 7.3 Final prominence vector and relationship vector for DEMATEL.

i/j	Row sum vector (Ri)	Column sum vector (Cj)	Relation vector R_i-C_j	Prominence vector R_i+C_j	Relationship between criteria
1	1.43189423	2.01246749	-0.5805733	3.4443617	effect
2	2.18718645	1.11601161	1.07117484	3.3031981	Cause
3	2.37938662	0.72749205	1.65189456	3.1068787	Cause
4	1.4764415	0.69863472	0.77780678	2.1750762	Cause
5	1.75218619	1.16680301	0.58538318	2.9189892	Cause
6	1.35219403	2.35059274	-0.9983987	3.7027868	effect
7	1.42565207	2.53821647	-1.1125644	3.9638685	effect
8	1.17946344	2.2077849	-1.0283215	3.3872483	effect
9	1.81807685	2.12645337	-0.3083765	3.9445302	effect
10	1.09920642	0.91925671	0.17994971	2.0184631	Cause
11	2.46853037	1.72032348	0.74820689	4.1888539	Cause
12	2.30258823	1.22289299	1.07969524	3.5254812	Cause
13	1.97239556	2.11926616	-0.1468706	4.0916617	effect
14	1.73356767	1.67864141	0.05492626	3.4122091	Cause
15	0.67323195	1.36736562	-0.6941337	2.0405976	effect
16	1.73815641	2.33829521	-0.6001388	4.0764516	effect
17	2.15569042	1.74220996	0.41348045	3.8979004	Cause
18	2.61021609	1.9727395	0.63747659	4.5829556	Cause
19	1.70558105	2.07383459	-0.3682535	3.7794156	effect
20	1.31741513	1.41784687	-0.1004317	2.735262	effect
21	0.78997934	2.05191116	-1.2619318	2.8418905	effect

When the criteria are grouped into the cause-and-effect group, we are able to establish which criteria need to be looked into in order to get a better output. Here, the better output is a better education supply chain. The service providers need to be trained properly such that they are able to impart good level of knowledge to their students. The cause-and-effect groups tell us that the teacher trainers need to concentrate on the cause group while developing their future training so that they can develop a good online training. The effect groups will be enhanced when the cause group is taken care of while developing the online training. If there are negative effects of certain causes then they need to be minimized and if there are positive effects, they need to be further enhanced so that eventually the online training is a good output. This good output of the education service supply chain would in turn a better employable workforce. Figure 7.4 shows the diagraph, which shows the cause and groups divided as per the relationship and prominence values as given in Table 7.3.

Figure 7.3 is the diagraph. From the graph, we can see the two groups viz. the cause-and-effect group distinctly as marked by the different colored outlines. We can see that criteria 2, 3, 4, 5, 10, 11, 12, 14, 17, and 18 are part of the cause group, and criteria 1, 6, 7, 8, 9, 13, 15, 16, 19, 20, 21 are part of the effect group.

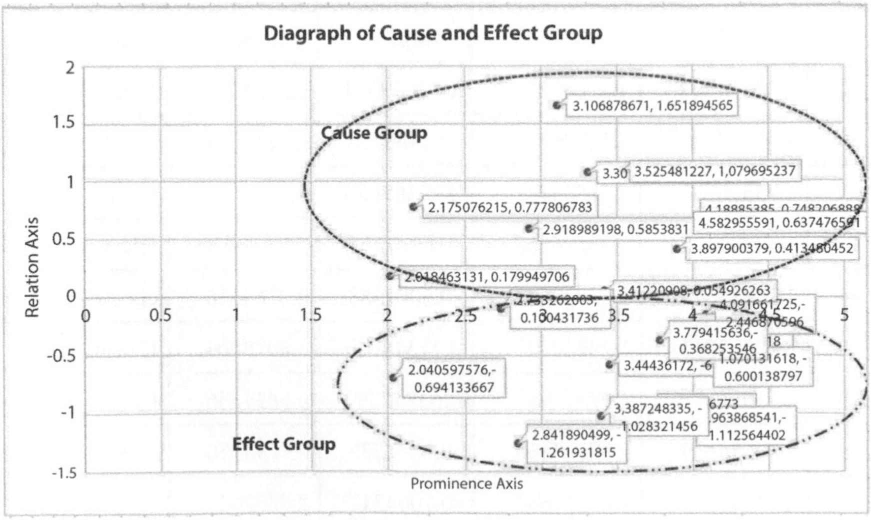

Figure 7.4 Diagraph of cause-and-effect group from DEMATEL.

7.6 Conclusion

The aim of the current research was to consider the education supply chain from the perspective of a teacher trainer in the context of online training in order to eventually develop a connected smart city environment through education. The pandemic has compelled everyone to consider how to change their style of functioning in order to be able to give the required output. It is understood that the pandemic has changed the way the entire world works and will work for a while. It is imperative thus, to ensure that we give it our best shot so that good outputs can be achieved. In a service supply chain, the good output is good service. In this chapter, we consider the evaluation of the criteria for developing online training in context of the teacher trainers. The evaluation is carried out using DEMATEL, which is a MCDM technique to evaluate conflicting criteria. When we are able to evaluate the criteria, we are able to establish the cause-and-effect group. In order to achieve a good output of the educational service supply chain, we are able to keep a check on the causes of the system, which would lead to better effect on the overall service output. The limitation of the study that the supply chain has considered is only the perspective of the service supply chain, whereas there is an impact of all the other stakeholders also on the entire educational supply chain. In the future, an expanded study of the overall educational supply chain in the context of technology usage for smart city implementation can be developed.

References

1. Towill, D.R., Supply chain dynamics. *Int. J. Comput. Integr. Manuf.*, 4, 4, 197–208, 1991.
2. Lee, H.L. and Billington, C., Managing supply chain inventory-pitfalls and opportunities. *MIT Sloan Manage. Rev.*, 33, 3, 65–73, 1992.
3. Davis, T., Effective supply chain management. *MIT Sloan Manage. Rev.*, 34, 4, 35–46, 1993.
4. Simchi-Levi, D., Kaminsky, P., Simchi-Levi, E., *Designing and Managing the Supply Chain: Concepts, Strategies, and Case Studies*, McGraw-Hill/Irwin, Boston, 2008.
5. Lambert, D.M., Cooper, M.C., Pagh, J.D., Supply chain management: Implementation issues and research opportunities. *Int. J. Logist. Manage.*, 9, 2, 1–20, 1998.
6. Chase, R.B., Jacobs, F.R., Aquilano, N.J., *Operations Management for Competitive Advantage*, McGraw-Hill Higher Education, New York, 2005.

7. Chopra, S. and Meindl, P., *Supply Chain Management: Strategy, Planning, and Operation*, Pearson Prentice Hall, Upper Saddle River, 2007.
8. Quinn, J.B., *Intelligent Enterprise: A Knowledge and Service Based Paradigm for Industry*, Free Press, New York, 1992.
9. Machuca, J.A.D., de González-Zamora, M.M., Aguilar-Escobar, V.G., Service operations management research. *J. Oper. Manage.*, 25, 3, 585–603, 2007.
10. Sampson, S.E., *Understanding Service Businesses: Applying Principles of the Unified Services Theory*, 2nd ed, John Wiley & Sons, New York, 2001.
11. Sampson, S.E. and Froehle, C.M., Foundations and implications of a proposed unified services theory. *Prod. Oper. Manage.*, 15, 2, 329–343, 2006.
12. Habib, M.M. and Jungthirapanich, C., An integrated framework for research and education supply chain for the universities, in: *2008 4th IEEE International Conference on Management of Innovation and Technology*, IEEE, pp. 1027–1032.
13. Ellram, L.M., Tate, W.L., Billington, C., Understanding and managing the services supply chain. *J. Supply Chain Manag.: A Global Rev. Purch. Supply*, 40, 4, 17–32, 2004.
14. Baltacioglu, T., Ada, E., Kaplan, M.D., Yurt, O., Kaplan, Y.C., A new framework for service supply chains. *Serv. Ind. J.*, 27, 2, 105–124, 2007.
15. Bolliger, D.U. and Wasilik, O., Factors influencing faculty satisfaction with online teaching and learning in higher education. *Distance Educ.*, 30, 1, 103–116, 2009.
16. Nie, W. and Kellogg, D.L., How professors of operations management view service operations. *Prod. Oper. Manage.*, 8, 3, 339–355, 1999.
17. Normann, R., *Reframing Business: When the Map Changes the Landscape*, John Wiley & Sons, Hoboken, NJ, 2001.
18. Lovelock, C., Classifying services to gain strategic marketing insights. *J. Market.*, 47, 3, 9–20, Summer, 1983.
19. Grönroos, C., Service logic revisited: Who creates value? And who co-creates? *Eur. Bus. Rev.*, 20, 4, 298–314, 2008.
20. Curran, C., Online learning and the university, in: *Economics of Distance and Online Learning: Theory, Practice, and Research*, W.J. Bramble and S. Panda (Eds.), pp. 26–51, Routledge, New York, 2008.
21. Hogan, R.L. and McKnight, M.A., Exploring burnout among university online instructors: An initial investigation. *Internet High. Educ.*, 10, 2, 117–124, 2007.
22. Moore, M.G. and Kearsley, G., *Distance Education: A Systems View*, Wadsworth, Belmont, CA, 1996.
23. Allen, I.E. and Seaman, J., *Online Nation: Five Years of Growth in Online Learning*, Sloan-C, Needham, MA, October 2007, Accessed on 21st January, 2021 retrieved from http://www.sloanconsortium.org/publications/survey/pdf/online_nation.pdf.

24. Rovai, A.P., Ponton, M.K., Baker, J.D., *Distance Learning in Higher Education: A Programmatic Approach to Planning, Design, Instruction, Evaluation, and Accreditation*, Teacher's College Press, New York, 2008.
25. Matorera, D., A conceptual analysis of quality in quality function deployment-based contexts of higher education. *J. Educ. Pract.*, 6, 145–156, 2015.
26. Carver, C.L. and Katz, D.S., Teaching at the boundary of acceptable practice: What is a new teacher mentor to do? *J. Teach. Educ.*, 55, 5, 449–462, 2004. Accessed on 21st January, 2021 Retrieved from http://dx.doi.org/10.1177/0022487104269524.
27. McCaughtry, N., Martin, J., Hodges Kulinna, P., Cothran, D., What makes teacher professional development work? The influence of instructional resources on change in physical education. *J. Serv. Educ.*, 32, 2, 221–235, 2006.
28. Sagnak, M., Ada, N., Kazancoglu, Y., Tayaksi, C., Quality function deployment application for improving quality of education in business schools. *J. Educ. Bus.*, 92, 5, 230–237, 2017.
29. Shieh, J.I., Wu, H.H., Huang, K.K., A DEMATEL method in identifying key success factors of hospital service quality. *Knowl.-Based Syst.*, 23, 3, 277–282, 2010.
30. Harford, J., Hudson, B., Niemi, H., *Quality Assurance and Teacher Education: International Challenges and Expectations. Rethinking Education. Volume 6*, Peter Lang Ltd, International Academic Publishers, Evenlode Court, Main Road, Long Hanborough, GB-Whitney, Oxfordshire, OX29 8SZ, UK, 2012.
31. Kim, K.J. and Bonk, C.J., The future of online teaching and learning in higher education. *Educause Q.*, 29, 4, 22–30, 2006.
32. Sampson, S.E., Customer-supplier duality and bidirectional supply chains in service organizations. *Int. J. Serv. Ind. Manage.*, 11, 4, 348–364, 2000.

8

Industry 5.0: Coexistence of Humans and Machines

Sandesh Kumar Srivastava[1]*, Pallavi Goel[2], Anisha[2] and Savita Sindhu[2]

[1]Galgotias University, Greater Noida, Uttar Pradesh, India
[2]Manav Rachna International Institute of Research and Studies, Faridabad, India

Abstract

The fourth industrial revolution carries with its various expectations for the future of sustainable and effective development. Industry 4.0 takes a significant part of everyday life. The application of digital life on smartphones, touch screens, and smart classrooms may be significant indicators of this. Not only computer information and life but also human-like robotics would take a long time. In the immediate future, we will also see the picture of a modern 5.0 model for manufacturing. Because of emerging technology and increasingly evolving and increasing AI-based approaches, remaining on the top is getting more and more challenging. Robots are much more relevant as now the brain machine's technology can be merged with the human mind, as artificial intelligence advances. There are coercive problems in the world economy because efficiency is to be improved without the human workforce being robbed of the manufacturing sector. The chapter presents sustainable business 5.0, where machines are intertwined with the human brain and function as partners rather than rivals, against these obstacles. Besides, the effect of industry 5.0, which is more likely to create more jobs than to reduce, also discussed on manufacturing and the economy as a whole. The chapter describes modern technology, from IoT to news. The integration of these innovations would turn industry 4.0 into industry 5.0.

Keywords: Industry 4.0, industry 5.0, artificial intelligence, manufacturing, sustainability, human-machine coexistence, collaboration, IoT (Internet of Things)

**Corresponding author*: sandesh.srivastava7@gmail.com

Loveleen Gaur, Vernika Agarwal and Prasenjit Chatterjee (eds.) Decision Support Systems for Smart City Applications, (137–152) © 2023 Scrivener Publishing LLC

8.1 Introduction

The past has been marked by three industrial revolutions. In the opinion of many, we are at the beginning of a new industrial revolution. Industry 4.0 is the fourth transition. In 2011, the idea was introduced recently. Because industry 4.0 was declared as a core policy in the high technical strategy by the Germans in 2011, this subject has been discussed in various scholarly journals, realistic papers and conferences [3]. McKinsey's definition of industry 4.0 is like "the next phase in the digitization of the manufacturing sector, fueled by four disruptions: the staggering increase in data volumes, computational power, and connectivity, especially new low-power wide-area networks; the emergence of analytical and business intelligence capabilities; new forms of human-machine interaction, such as touch interfaces and augmented reality systems; and improvements in transferring digital instructions to the physical world, such as advanced robotics and 3D printing" [7]. Industry 4.0 has four main components: cyber-physical networks, the internet and smart plant. Machine-for-machine (M2M) and intelligent device modules in the industry 4.0 are not deemed separate since M2M is an Internet of Things enabler, while intelligent devices are a virtual physical network subcomponent. The fourth step of the digital revolution named the industry 4.0 virtual physical structure focuses on linking items across the Web in the supply chain. Internet networking innovations have allowed communication anywhere and anytime, increasing the importance of wireless links. Through the growth and transformation phase of industry 4.0, there are tremendous resources to support companies, communities and organizations [2]. There are six key innovations (IIoT and CPS, additive (3D) printing), large-scale technology (AIS), interactive robotics (CoBot) and virtual reality) that build industry 4.0. There are six main innovations to grow industry 4.0. Possible changes in the labor market due to industry 4.0 are only a result of the HR agreement. This is unacceptable and has been expressed in many industry 4.0 related papers. The future of work in industry 4.0 without humans is truly unimaginable. Market 4.0's goal does not depart from previous developments in the market. Its fact, with the aid of emerging technology, industrial manufacturing is accomplished. Engineering continues to establish the technical demands for manufacturing and markets. The center of industry 4.0 is smart growth. In this relation, the so-called revolt is a top-down strategy. You may claim that it often needs a bottom-up strategy. In the eyes of SMEs, Business 4.0 for starters, is meaningless. In several businesses in several business industries, the costs of adopting innovations, such as IoT, automation and big data, are heavy.

Due to expanded automation, more workers can be without jobs. Process automation in the manufacture of products forces companies to be connected to both customers and suppliers in both directions at all times.

Information technology is a platform for Web access (web, ERP and application), which enables consumer entry into the chain and remain linked. The order flow is led directly to the components needed for the request, which will be transported to the assembly plant, where the components for the assembly are tracked. The assembly consolidates the actual shipment sent by the customer during transit to the nation, indicating the shipping model of the final user (air, maritime or land).

A full process signal is produced by the existence of components and appropriate arrangement of things; the order needs the existence and properly installed signals to be delivered to the user. The Internet of Things (IoT), linking non-computers and non-human objects, regulating different facets of things, will exchange them with manufacturers, customers, institutions, and the community at large. There must be more new jobs than the losses of jobs. Otherwise, the current digital technology would be harmful to popular opinion.

This double contact, industry 4.0, is losing ground in industry 5.0. Since then, in this modern age, man has faced the computer. While industry 4.0 is still in the early stages and is projected to accomplish significant milestones only in 2020-2025, you see an image of a modern industry 5.0 model. Industry 5.0 is the business in which the state-of-the-art IT, IoT, robotics, artificial intelligence, AR technology, and other fields of operation, not for the sake of advancement, but the profit and comfort of each individual, are frequently used in their daily lives of human beings. After the 1960s, the usage of robots in engineering has grown as part of business 3.0 first. Robots developed in particular in the automotive industry where they have been primarily used to link bodies for welding processes. With the complexity of the demands, robotics was introduced in many fields such as manufacturing, pharmacy and the food industry. In 2006, more robots were used outside the car sector than within the sector for the first time. Currently, robots are found in small to medium-sized businesses not only in big factories but also in more inexpensive to simpler integrated robotics [5].

Driven by the need to manufacture at least new and inexpensive, high-quality goods, the bulk adoption of today's technologies is generally embraced by industry 4.0, which involves Internet links between consumer order networks, supply chains and even robots on the automotive production level [2]. On the other side, industry 5.0 products/services that collaborate between man and computer, enable citizens to express themselves through the simple human urge. The goods that bear the distinctive label of human treatment and crafts are customers that want the most and spend the highest. It is called personalization, in other terms. Industry 5.0 is also a transition to pre-industrial development; however, it is rendered feasible by the new technologies available.

This lesson shows the definition of industrial 5.0. The report further analyzes the transition from IoT to digital technology and the creation of companies where writers work, in which robots work with human brains and act as partners rather than rivals. We believe that the integration of such innovations will transform industry 4.0 into industry 5.0.

8.1.1 Industrial Revolutions

The origin of the word dates back to the 15th century. However, the first industrial revolution began in the late 18th century [1].

First industrial revolution (industry 1.0): The first industrial revolution in 1780 came in when wind, steam, and fossil fuel provided mechanical resources. The usage of advanced equipment and emerging technology developments demonstrates a large rise in inefficiency.

Second industrial revolution (industry 2.0): In the second industrial revolution, the factories of assembly lines and mass output preferred electric technology in the 1870s. In this technological transition, power was the primary source of resources.

Third industrial revolution (industry 3.0): In 1969, the third industrial revolution introduced the concept of automation to manufacturing industries. The development to the use of industrial processes for software and IT.

Fourth industrial revolution (industry 4.0): The fourth stage of transformation uses the Internet of Things (IoT) and cloud machines to provide real-time connections between so-called virtual and physical universes called cyber-physical networks. As in rising economic revolutions, the fourth revolution was fuelled by numerous technical advances.

Fifth industrial revolution (industry 5.0): Autonomous processing on and off the circuit with human intelligence and development of robot facilities

8.2 Literature Review

8.2.1 Industry 4.0 Characterization

The industrial revolution 4.0 was announced by the German government at the Hannover Messe Trade Fair in 2011 and (recognized) and describes the fourth stage of the industrial revolution through improvements in manufacturing information and communication technologies [14]. "It is the fourth stage, the industrial revolution, that is characterized by a high degree of automation, with automated equipment, management (certain cyber-physical systems, and the use of cloud technologies and big data". Industry 4.0 includes a wide range of concepts, such as digitization, automation, standardization, dynamic and secure network minimization, as well as step-by-step development and general innovation framework in this area [14]. Cyber-physical control systems for physical processes, real-time supply chains, and the Internet of Things (IoT) [14]. The most fundamental changes of industry 4.0 are based on research and development technologies, such as artificial intelligence, data analytics, Internet of Things, cloud technology, robotics, blockchain technology, 3D printing, cryptocurrency, and more [14]. Synthesizing from four principles specific to industry 4.0.

- Interoperability — Support for industrial machinery, equipment integration and machine-to-machine communication using IoT.
- Information transparency — The computer can create virtual copies of real-world objects.
- Technical support — A computer machine with AI to sup port the person, employees and effectively.
- Decentralizing the application of tech systems that can act as, and perform their functions. These ideologies should sustenance the transition scenario of industry 4.0. Successful reforms, the three-dimensional view you need [13].

(a) Horizontal integration, value creation, between organizations and enterprises raising the artefact life cycle and efficient financing, managing material flows, through real-time information exchange and careful arrangement. (b) Vertical integration: association across several hierarchical levels of the group based on interpersonal relationships and digitization and their divisions. This means that the company's transition to an intelligent system is a flexible exchange of information in real-time and careful planning. (c) End-to-end design, design and development of products, and facilities tailored to the requirements and needs of customers using digital technologies. Currently aimed at attracting customers, it is based on finding the best products and services (it is on-demand) and meeting their needs. This, combined with the constant improvement of the company's business processes, is in theory cheap, quality of production, receipt.

8.2.2 Definition of Digitalization

The term "digitization" is used to refer to inflation and many different signals, including the following examples: digital transformation, digital business, digital society and the digital revolution. In the early literature, digitization was the definition of the transfer of analogue information to digital data, and consequently, referred to as digitization. Digitization is the next step in transforming enterprise information into electronic processes, converting electronic documents into a single digital management format and thereby increasing efficiency, flexibility and cost, saving process and bringing them to market in less time [15]. For businesses, described digitalization transformation, digital skills in society, work experience, and related variations in relationships, in organizations and on persons. Urbach describes the reduced time to market, as well as the commercialization of technologies, as the driving force behind digitization [16]. In addition, in digital technologies, there are, on the one hand, certain methods, such as social media, mobile calculations, advanced analytical, and cloud computing (SMAC) [17], and on the other hand, new know-hows, such as artificial intelligence, blockchain, and the Internet of Things. Time and place, regardless of the availability of information, there are advantages of digitalization. This will ensure the previously unseen rate of change and degree of connectivity of the entire supply chain, which encircles the dominant role of the consumer [16].

8.3 Requirement of Fifth Industrial Revolution

Industry 4.0 is for machine management and the deployment of edge computing in a smart way. The emphasis is primarily on increasing production performance and unintentionally overlooking the human costs arising from production optimization. As a result of the demand for better employment, it would encounter opposition on the part of the trade groups and lawmakers that would represent industry 4.0. Industry 6.0 is suggested as the approach to sustainability that is required.

Furthermore, after the Second Industrial Revolution, the planet has undergone a huge rise in environmental emissions. Sadly, industry 4.0, while several various AI systems have been used to carry out work from an ecological viewpoint, has not been heavily based upon environmental protection nor has it concentrated on the technology that enhances Earth's environmental sustainability, the past 5 years, the previous 50 years.

While industry 4.0's biggest problem is robotics, industry 5.0 would be a partnership of people and self-employed computers. Robots will not only be a programmable computer but also still be a perfect human partner to execute routine activities. The new technological revolution is going to launch the next wave of robots that are usually referred to as cobots. Such interactive robots are mindful of human interaction and are also liable for the health and danger criteria.

Advanced technologies required [4]:

- A sensor network with low intelligence and processing capacity could minimize the need for a data transmission system with high bandwidth while enabling some local data preprocessing. Besides, this will reduce the cost and latency of the network, thus providing "distributed information" on the network.
- Through integrating virtual and physical environments, manufacturers can evaluate data, track the production cycle, mitigate hazards until they emerge, decrease downtime, and evolve more by simulations. Now it is possible to build much more accurate digital twins, which better model the complexity of circumstances and operational characteristics for an operation, with advances in big data analytics and artificial intelligence.
- Floor trackers improve real-time output monitoring. They allow customer selling orders to be coupled with production orders and additional materials. They allow asset and process flow analysis in real time, paving the way for an integration of the online process in the production process. These trackers may be used as network sensors or use the advantages of network sensors.
- Cost and resources with all sides were greatly reduced. Digital education is essential for having a skilled workforce that does not jeopardize the efficiency of managing or place a human worker at risk. For occupations and activities that involve a form of danger attributable to repetitive activity or behavior at work, it is especially important.

- Machines can practice AI techniques and therefore carry out the intended function autonomously. Classification, regression and clustering of deep learning methodologies contribute to clever structures and algorithms, which can make unpredictable decisions. This also transmits the digital/virtual system's gained skills and experience to its sibling, who play a significant role in the 5th industrial Revolution.
- When a human being thinks something abnormal is in his office, he will stop functioning even though nothing is wrong with his eyesight or his emotional intelligence. That kind of preparation is necessary to avoid injuries at the workplace. This cannot be done at this point with our vision and perception of technologies. Beyond visual and tactile equipment, the computer has to develop awareness to determine the best in the ever-changing workplace.

8.4 Journey of Industry 4.0 to Industry 5.0

There is a seven-step convention for convergence of science and technologies in society [7]:

Step 1: Management in the era of self-organization,
Step 2: Emergent intelligence,
Step 3: Theory of complex adaptive systems,
Step 4: Ontology and knowledge basis,
Step 5: Multiagent systems in digital eco-systems,
Step 6: Internet of everything,
Step 7: New types of distributed computers and distributed robotics.

The philosophy of intersubjective management processes where any involved agent will show himself to be the universal "user," absorbed in a certain dilemma and ready to be willing to engage with the other actors in his response, is at the heart of management throughout the era of autonomy.

Emerging knowledge is an unusual trend in which large entities emerge from encounters with small or simplified entities, such that larger entities show features that are not smaller or simple ones.

Simple adaptive systems should be used as a foundation for multiagent schemes for structured adaptive schemes. If answers to a complicated problem are found through self-organization, it integrates multiagency processes with nonlinear thermodynamics.

For multiagent structures, the ontological paradigm was commonly embraced, where ontologies are both certain information bases of theoretical agents comprising both the awareness of one specific data area and the information of the decision-making processes.

The multiagent framework is characterized as a network of slightly linked private problem solvers (agents) that exist and interact in the general environment to achieve these or other framework purposes. The interaction may be done by direct paths, exchanging of messages or implicitly because certain agents take into consideration the existence of other agents by modifying the external atmosphere in which they communicate.

To support conventional and common citizens, the IoT is evolving intensively innovations which are the base for automation in industry 4.0 and 5.0.

Computer networks spread to provide a hierarchical topological network layout and include machine computing with concurrent and asynchronous processes [12].

8.4.1 Industry 4.0

Producers contend for rising business requirements. This calls for versatile, knowledgeable, and scalable manufacturing lines to fulfil current requirements. Business executives and procurement administrators also agreed that corporate and industrial development convergence will be accomplished. The convergence of various facets of a business, including manufacturers, manufacturing lines, and consumers is the best way to do this. The Internet of Things (IoT), which is the key tool of industry 4.0, was named this multifaceted incorporation. The user interface monitors the communication mechanism and all the modules send out signals for the planned tracking which shows progress on the device, so users depend on machine details based on the IoT communication protocol.

IoT paradigms involve the human-machine interface, somaticized paradigms that maximize connectivity importance to human beings [2]. Industry 4.0 of the Internet of Things will supply the management process with data and knowledge. The process for communicating between items speeds up the numerous phases in the supply chain. An order for a variety of components in various suppliers that can be linked anywhere on the Internet, provide detailed details regarding the product and track them throughout the logistics process.

All pieces produce a key signal to complete the order, following the internet monitoring program, to hit the production location. The entire order is fallow for the device and sends signals in the back-end windows to obey the order method. The transparent and centralized design of IoT will require peer analysis of the broad data gathered by individual users and scientists in group environments for triangulation and testing. In addition to making a sense of big data, doctors, the public and informed citizens should have more "after" incentives to relate to the study design and implementation of the study agenda [6]. Industry 4.0 is based on cyber-physical systems (CPS) which communicate via IoT. Efficient and safe storage processes are essential for the sharing of information in real-time between CPS. The most popular approach is cloud computing. Several analyses and procedures are often important to extract valuable knowledge from broad and raw data lakes. The next development to merge the technical and physical realms was the Global Internet that integrated the examined data with the IoT [4].

Industry 4.0 incorporates IoT in workstations and examines the captured big data in the cloud to improve flexibility and safety efficiently. The ecosystems of the invention must be regulated, and cannot be left alone. Decisions about the choice of philosophical frameworks (epistemologies) which guide innovation ecosystem governance are critical since they affect what, when, when, how and for whom or for whom innovations materialize. The epicenter of health and life sciences has still not yet been fully founded in industry 4.0. Also, IoT and industry 4.0 are differentiated in many fields such as AI, aerospace and automated vehicles, industrial infrastructure, supply chain management, logistics, consumer relations, and retail [6].

8.4.2 Industry 5.0

For Business 5.0, there are multiple dreams. While industry 4.0 involves linking computers, industry 5.0 claimed that it was about communication between person and computer. Instead of regular robotic development, an innovative human touch should remain. This is going to build fresh work. Employees of the company should play different positions. Robotics and AI are some of the cornerstones of industry 5.0. Human-machine communication is a common continuation of robotics and AI.

Industry 5.0 is an incremental (but essential) development that shapes the definition and observes of industry 4.0. To this end, it is an essential goal of industry 5.0 to tackle the four historically overlooked asymmetries in the architecture of the 4.0 environment, within creative global governance structures. Many more would prefer to call the mark industry 4.0 plus, industry 4.0 Symmetric, industry 4.0-S, among others, because the architecture of the technology ecosystem industry 4.0 is focused on previous, theoretically defective holes among asymmetries [6].

For example, the human starts a mission, and the robot uses a rotor camera to track the process. It functions as the head of the computer. The robot is also linked through machine learning to a computing device that takes the photo, processes the image and learns patterns. We also track the person, control the world and predict, through a profound psychological intention-driven examination, what the user may do next. If the robot learns its forecast, the human worker may seek to support.

The usage of the human brain as a light source is another form of intelligent identification.

Electroencephalography (EEG), infrared spectroscopy (fNIRS), may be used for this method. These fNIRS headphones monitor brain activity efficiently and are ideal for a broad spectrum of activities including signal processing, goal predicting and background perception. These fNIRS may be used to carry out a specific function in a medical centre where the operator may monitor a robotic arm with a diagnostic or surgical instrument [4].

8.5 Industrial Revolution: Changes and Advancements

8.5.1 Big Data

Big Data needs improvements in the way we interpret big data in the flow due to its immense scale, pace and range, not to mention the contested veracity [6]. Real-time processing is needed by Big Data to grasp and catch the meaning of the time, place, and structure of its transparent type and its continuously shifting dynamic existence. In brief, exploration research is not what it used to be in the internet and big data age. There might well not be a strong example of old methods of doing research, but in the field of digital data application technology and the complexities of big data in health, in particular, the paths forward are not obvious. Those who talk nice regarding proactive creativity. Big technology is still at the moment of engagement as Big Data and its transparent nature and features are not adequate and inadequate to reflex the pattern of partnership building and related concrete approaches to turn developments into global progress.

Instead, the big data deployment science mentioned above requires Cyber Physics Systems (CPS).

Large-scale technology like Big Data in healthcare needs not only large-scale manufacturing but custom-designed development in smart plants, such as precision medicine. Therefore, engineering architecture is a pillar of medical device research. The characteristics of real-time manufacturing versatility for differentiation vs mass production of health technologies would have to be discussed by IoT, AI or any platform which can be used to turn big data into knowledge-based innovation.

8.5.2 Changes the Fifth Industry Revolution Might Bring

Industry 5.0, as machinery tightly interacts with a human being's everyday existence, should face unparalleled challenges in human-machine interaction (HMI). Industry 5.0 will revolutionize manufacturing processes globally by the removal of human employees 'tedious, filthy and repetitive jobs, wherever feasible. Robots and autonomous devices can hit an unparalleled stage of output supply chains and manufacturing plants. Industry 5.0 is planned to improve efficiency and organizational efficacy, to be eco-friendly, to minimize worker risk and to shorten development times. It generates more workers than it destroys. Several innovative companies are building modern platforms to offer unique hardware and software applications worldwide. It would improve the global economy and increase the world's cash flow [4].

8.5.3 Working with Robots

The man knows robots as an idea from books and movies. However, the human-robot coworker is new to almost all employees in the office. Although it is easy to communicate with a robot, the experience is different. We need to learn how to deal with robots to get what we want. Informal communication is an important part of human interaction. Robots may not be able to understand this simple conversation. Humans need time to learn and learn to work with robots. In addition, there are robots with different abilities. Some thrive, some do not. Separating robots with different abilities can be difficult and confusing. Robot manufacturers need to find a way to develop robots that tell their capabilities without any confusion. Humans need to learn to work with different types of robots [8, 11].

8.5.4 Impact of Fifth Industrial Revolution on Education

With the growth of the industry in the fifth industrial revolution, the education sector is one of the most affected. In terms of new technology booming, we are stirring to the fifth educational revolution or Education 5.0. The evolution of the industry has removed almost all the barriers in the education sector and everyone is about exploring new technologies and creating a better among the beings.

8.5.5 Problems which the Fifth Industry Revolution Might Generate

People can attach value-added tasks to their development strategies, causing significant shortages in skills and becoming the key problem for the high workforce of the big employment created. New industrial development would find change even easier for the older population of an organization and stakeholders. Quick and highly efficient manufacturing can create overproduction problems.

The problem of adaptation to the human-machine relationship cannot be ignored. Man identifies robots as a concept from books and movies. However, human-robot colleagues are new to almost all employees in the workplace. Although it is thought to be easy to communicate with robots, the experience may be different than expected. We need to learn how to act towards robots to get the job done. Nonverbal communication is an important part of human interaction. Robots may not understand this nonverbal communication. Humans need time to work and learn with robots. Also, there are robots with different abilities. Some thrive and some do not. Separating robots with different abilities can be difficult and confusing. Robot manufacturers need to find a way to develop robots that demonstrate their capabilities without any illusions. Humans need to learn to work with different types of robots [9].

8.5.6 What Might Help to Solve the Problems

Uniformity along with regulation would help to escape severe scientific, social and industry issues. Implementing accountability always needs to be addressed. They ought to understand how autonomous institutions will incorporate concepts of ethics. In autonomous systems, ethics approaches can be clarified to test and confirm ethical behaviour, even in autonomous systems.

8.5.7 The Role of Ethics in Industry 5.0

As companies see increasing levels of human-machine collaboration, further ethical questions and concerns are being raised about the impact of technology on the engineering of future industrial systems [10]. We argue that 5.0 ethics in the industry promotes a natural relationship between human beings and the cyber-physical realm world, that justice serves as an autonomous mechanism for balancing human interests and for the well-being of society. Human-robot co-work problems have been identified, for example, mental health problems, lack of social interaction, scepticism about robot learning, declining human workforce, and human-robot competition [1]. Although this research is very valuable in bringing forward specific values, it generally does not take a holistic approach. This is especially true when deer are competing. Thus, although the values and face-to-face interviews that affect their relationship with the theoretical background and technology for the future of industrial workers are widely accepted, current research in this area is very limited, and most research efforts emphasize equally the technical aspects of the future place of work.

8.6 Conclusion

Studies indicate that the growing various realistic implementations and the rising growth of the digital economy are now creating a solid base for the advancement of Industry 4.0 technology and that this may function as a launchpad for industry 5.0 in the long term. Such visionaries often demonstrate industry 4.0 vulnerabilities and recommend industry 5.0 to fix industry 4.0 limitations. It is premature and suggested without sophistication at the outset of artificial revolution. They will witness a systemic transition in all markets, businesses, and cultures to term a phenomenon an industrial revolution. The key topic of the 5.0 guidelines is sustainability. Mass development and productivity are not exclusively related. Therefore, it may be a smarter strategy to merge these two aims or principles and redefine the next technological revolution. The fifth industrial revolution would occur as the technological environment in conjunction with human intellect completely combined the three main components—intelligent machines, intelligent networks, and smart automation. The faith and trustworthiness between the two pieces provide promising performance, seamless quality, reduced waste, and flexible output. You would, therefore, draw more customers to the place of work and improve service performance.

References

1. Demir, K.A. and Cicibas, H., Industry 5.0 and a critique of industry 4.0. *4th International Management Information Systems Conference*, Istanbul, Turkey, 2017.
2. H., A. and P. Cisneros, M.A., Industry 4.0 & internet of things in supply chain, in: *Proceedings of the 8th Latin American Conference on Human-Computer Interaction (CLIHC '17)*, Association for Computing Machinery, New York, NY, USA, pp. Article 23, 1–4, 2017, https://doi.org/10.1145/3151470.3156646.
3. Hermann, M., Pentek, T., Otto, B., Design principles for industrie 4.0 scenarios. *2016 49th Hawaii International Conference on System Sciences (HICSS)*, IEEE, 2016.
4. Nahavandi, S., Industry 5.0—A human-centric solution. *Sustainability*, 11, 16, 4371, 2019.
5. Ozkeser, B., Lean innovation approach in industry 5.0. *The Eurasia Proceedings of Science Technology Engineering and Mathematics*, vol. 2, pp. 422–428, 2018.
6. Özdemir, V. and Hekim, N., Birth of industry 5.0: Making sense of big data with artificial intelligence, "the internet of things" and next-generation technology policy. *Omics*, 22, 1, 65–76, 2018.
7. Skobelev, P.O. and Yu Borovik, S., On the way from industry 4.0 to industry 5.0: From digital manufacturing to digital society. *J. Ind. 4.0*, 2, 6, 307–311, 2017.
8. Demir, K.A., Döven, G., Sezen, B., Industry 5.0 and human-robot co-working. *Procedia Comput. Sci.*, 158, 688–695, 2019.
9. Longo, F., Padovano, A., Umbrello, S., Value-oriented and ethical technology engineering in industry 5.0: A human-centric perspective for the design of the factory of the future. *Appl. Sci.*, 10, 12, 4182, 2020.
10. Pacaux-Lemoine, M.-P. and Trentesaux, D., Ethical risks of human-machine symbiosis in industry 4.0: Insights from the human-machine cooperation approach. *IFAC-PapersOnLine*, 52, 19, 19–24, 2019.
11. Welfare, K.S. *et al.*, Consider the human work experience when integrating robotics in the workplace. *2019 14th ACM/IEEE International Conference on Human-Robot Interaction (HRI)*, IEEE, 2019.
12. Doyle-Kent, M. and Kopacek, P., Industry 5.0: Is the manufacturing industry on the cusp of a new revolution?, in: *Proceedings of the International Symposium for Production Research 2019. ISPR 2019, ISPR 2019, Lecture Notes in Mechanical Engineering*, Springer, Cham, 2020 https://doi.org/10.1007/978-3-030-31343-2_38.
13. Paschek, D., Mocan, A., Draghici, A., Industry 5.0-The expected impact of next industrial revolution, in: *Thriving on Future Education, Industry, Business, and Society, Proceedings of the MakeLearn and TIIM International Conference*, Piran, Slovenia, 2019.

14. Ustundag, A. and Cevican, E., *Industry 4.0: Managing The Digital Transformation*, Switzerland Springer Nature Switzerland AG, Cham, 2018.
15. Köhler-Schute, C., *Digitalisierung und Transformation in Unternehmen: Strategien und Konzepte, Methoden und Technologien, Praxisbeispiele*, KS-Energy-Verlag, Berlin, Germany, 2016.
16. Urbach, N., *Digitalization Cases, How Organizations Rethink Their Business for the Digital Age*, Switzerland Springer Nature Switzerland AG, Cham, 2018.
17. Paschek, D. *et al.*, Business process as a service-a flexible approach for it service management and business process outsourcing. *Management Challanges in a Network Economy, Proceedings of the Make Learn Conference 2017*, 2017.

9

Smart Child Safety Framework Using Internet of Things

Afzal Hussain[1], Anisha[2*], Pallavi Murghai Goel[2] and Sudeshna Chakraborty[3]

[1]*Galgotias University, Greater Noida, India*
[2]*Manav Rachna International Institute of Research and Studies, Faridabad, India*
[3]*Lloyd Institute of Engineering and Technology, Faridabad, India*

Abstract
The safety of a child is an important concern for every parent. The busy lifestyle restricts close monitoring of their children. The emphasis is on using enhanced technology to address this issue. Therefore, this paper provides an overview of how IoT technologies are efficient in developing a real-time monitoring application. The aim is to develop a new framework that monitors the movements of children and can provide an accurate location with the help of smart devices and different localization techniques. IoT emerges as an advanced automation and analytics system. It exploits sensing, networking, data, and different technologies to provide the whole system as a product. The proposed model will ensure the safety of children by integrating cell phones, tracking mechanisms, and cloud technologies. The system will keep track of the children and send real-time notifications to parents so that appropriate action can be taken in emergency situations. Also, certain sensors can be deployed, such as sensors for detecting body temperature and heartbeat to make sure the safety of a child. Hence, this chapter aims at providing parents with a sense of security for their child in today's era.

Keywords: Internet of Things, wireless sensor networks, child safety, security, GPS, user application

**Corresponding author*: anishanagpal@outlook.com

Loveleen Gaur, Vernika Agarwal and Prasenjit Chatterjee (eds.) Decision Support Systems for Smart City Applications, (153–166) © 2023 Scrivener Publishing LLC

9.1 Introduction

Internet of Things is defined as an interconnection of various computing devices, sensors, hardware, actuators having the ability to transfer data over a network. IoT is used in various applications such as home automation systems, agriculture, smart vehicles, smart grid, e-healthcare, etc [1, 2]. Therefore, IoT has emerged as a beneficial technology in providing smartness to a particular system.

In today's environment the security of a child is a major concern as child abduction cases are increasing day by day. The data released by NCRB (2016), India [3], reveals that there is an increase in crime rates against children by 20% in 2016-2017 and it increases over six times in a decade of 2008 to 2018. This shows that the primary concern is to save the child from being abducted or abused. The busy lifestyle of parents and lack of self-protection skill in children needed a smart tracking and monitoring system for safety.

The aim of this model is to provide a solution to this problem using Internet of Things and cloud technology. This proposed work not only focuses on the movement of a child outside the house premises, it also keeps check on the vital parameters of the human body by using sensors. It helps parents to monitor their child and keep track all the activities they are doing in real time without any human intervention.

The framework developed also helps in providing accuracy in location of child by using smart devices and localization techniques [4].

This paper is organized as Section 9.2 consists of the literature review, Section 9.3 introduces the technology and sensors used to solve the problem, Section 9.4 presents the Proposed system architecture and its working principle, Section 9.5 discusses the advantages of the proposed solution, Section 9.6 is the conclusion of the whole paper and Section 9.7 presents our future work.

9.1.1 Literature Background

There are various proposals regarding child safety has been given literature.

Moodbidri [5] proposes a smart wearable device for safety of a child. It can be used with any mobile phone and there is no necessary need for having smart phones. The aim of the device is to keep track of the child's location and sending alert messages using the GSM module. This device is having GPS location sensor, temperature sensor, UV sensor, SOS light and an alarm buzzer. Therefore, it enables the parents to have real time information about the child's whereabouts and acts as a smart IoT device.

An IoT-based hybrid model for monitoring child safety is proposed by R. Kamalraj [6]. It gives alert when someone will go close to the child and tracks the current location of the child. This model uses alcohol and gas sensors to check the surrounding conditions. It also uses the blood pressure sensor to check the child vital parameters. After processing the data gathered a message or alert is send to the mobile phone of guardian of the child.

A wristband named Vital having multiple sensors is proposed by Kyle Braam [7]. It is a low cost and light- weighted device for emergency and uses Arduino Microsystem and Bluetooth 4.1 module. It takes sample inputs from multiple sensors and sends response to the parent's mobile phone. The estimated running time is 60 or more hours as per the use cases. Also, different power consumption techniques may further reduce the consumption and can extend lifetime of battery up to 100 hours.

Ushashi Chaudhary [8] proposes a multisensory wearable device to track the safety and security of the child. This device can detect the child's temperature, surrounding humidity and heartbeat of child. It consists of SOS light and buzzer to notify bystanders about the emergency situation. Its main focus is on the communication through text SMS in which parents sends SMS having some keywords and the device reply back.

An IoT-based smart ring is proposed by Navya R Sogi [9] specifically for woman safety. This device helps in tracking location of women using GPS and also it captures image of the attacker through raspberry pi camera. It sends the current location of the person along with the image URL to the predefined mobile numbers and police by using smart phone.

An IoT-based framework is developed for the safety of children [10]. This application keeps track of a child from boarding a bus to entering into the classroom. In an emergency situation child can send signal to parents and school in order to alert them. This framework is specifically helpful for ensuring safety of children in school bus and its premises.

Qureshi [11] proposes the solution to child safety by using Wireless technology. The smart wearable device helps the parent of a child to accurately locate the child's movements and the surrounding temperature. This wearable device is based on Internet of Things, the parent need to send some specific keywords in order to get the information regarding child location. The drawback of this system is that it relies on the keywords to get the exact location.

Madhuri [12] provides an architecture which integrates IoT, cloud computing and GPS technologies to get the exact geographical location of the child. The proposed architecture takes four elements such as people, Information, process and digital technology. This system works only for tracking the movement of the child and not the surrounding environment.

An IoT-based wearable device is proposed by M. Pramod [13] for the safety of women. It examines the physiological signs along with the body gestures of the girl child or women. It uses the galvanic skin resistance to measure the body temperature. Therefore, it basically deals with the relationship between stress, body temperature and skin resistance.

Bhanupriya and Sundarajan [14] propose a wearable device that is integrated with multiple computing devices. It consists of wearable "Activity Tracker Wrist Band" with all the necessary information such as human reaction to anger, anxiety and fear. When any abnormal situation occurs and any of the reaction takes place, various sensors generate emergency signals that will be transmitted to the Smartphone. The system effectively monitors the presence of children within expected locations. When a person crosses a monitoring threshold, GSM sends a request for help by sending messages to the nearest police station, to parents and people around it.

9.2 Technology and Sensors Used

This section gives the brief about the technology and sensors used in the designing of this system.

1. Internet of Things: Internet of Things (IoT) is an interconnected system containing various computing devices, objects (things), softwares, and sensors that communicate with each other without human intervention using network technologies. IoT provides a platform where smart services are delivered to the users by connecting people and things that makes their life easy. It is used in various application domains such as agriculture, smart homes, healthcare, industries, and smart cities where IoT contributes in provisioning smart services [15, 16].

2. Cloud Computing: As we know, IoT produces a huge amount of data and this data needs to be stored, analyzed in order to have maximum advantage of an IoT system. Cloud computing not only provides a way to store and manage this data; it also helps in modernizing the operations. According to National Institute of Standards and Technology [17]: "Cloud computing is a model for enabling ubiquitous, convenient, on-demand network access to a shared pool of configurable computing resources (e.g., networks, servers, storage, applications, and services) that can be rapidly provisioned and released with minimal management effort or service provider interaction." IoT and Cloud computing both shares complementary relation and contributes in increase of efficiency. Apart from the various benefits, some particular issues can be seen in the service models of cloud

computing which are related to data integrity, network security, data confidentiality and service level agreements [18].

3. Wireless sensor Networks: It is the spatial distribution of sensors which monitors the environmental conditions such as temperature, pressure, sound, vibration etc. It takes the data and transfers it to the network. The sensor nodes in an environment communicate with each other wirelessly with higher accuracy. The application areas of WSN are health monitoring, environmental exploration, safety and security where timely data transfer is required [19].

4. Bolt Wi-Fi module: It is a small computer containing system on chip. It includes multi-core processor, input/output peripherals, GPU and ROM. It is useful for embedded, as well as IoT-based application. It consists of onboard Wi-Fi through which it can directly connect to the internet and send sensor data and receives data via a web application through internet [20].

5. Pulse Rate Sensor: It is used to measure the pulse or heart rate of an individual. It is based on photo phlethysmography principle. It is having a light emitting diode and a light detecting photodiode. It tracks the variation in the blood flow in the different region of body and change in the light intensity [21].

6. Temperature Sensor and humidity sensor: It is used to measure the body temperature of a person and humidity to check the physical condition. It requires RTD to detect the temperature. It can sense the heat index, body temperature and internal body humidity. We can use this to measure the child's temperature and detects the relative surrounding issues [22].

7. LED Light: It ensures to alert the bystanders by flashing the light. Whenever the value of this sensor crosses the threshold, the light blinks. It nowadays becomes a symbol for help and through this people can analyze that child is in need of help. It works on Morse code principle and parents can activate it by sending SOS message to the device. It helps all security personals to track the missing child [23].

8. Alarm Buzzer (Piezo): The buzzer is used to make alert tones and beep alarms. It alerts the bystanders when the sensor value reaches the threshold to specify that child is in emergency situation [24].

9. GPS: Global Positioning system is a navigation system used to track the real time location. It continuously takes data from satellite. Therefore, it is helpful in the applications such as fleet management, navigation, tracking system, mapping and robotics. GPS location sensor is used here to track

the child's location as it returns the precise latitude and longitude coordinates [25].

10. GSM: It is used to transfer the message from an individual to concerned authorities. It establishes the communication between a GSM, a computer and GPS. Here we are using GSM LT-2, which enables to give messaging and monitoring facilities in alarm systems [26].

9.2.1 Proposed System Model and Working

The proposed model of the system is based on IoT, and it needs no human intervention. The actions can be taken autonomously. The conceptual architecture of the system is shown in Figure 9.1.

Working Principle:
In the proposed model, pulse rate sensor and temperature sensor are used to check the vital parameters of a child in an abnormal situation. If the values are increased beyond the threshold value, then it raises an alarm. In the next phase, it tracks the location of child movement from the GPS attached to it and it directs the parents and concerned authorities for whereabouts of the child. All the information can be monitored from a mobile phone using an android application. Through this monitoring activity, parents can protect their children by the abusers.

Here, cloud server is used to store the crucial data for the functioning of the entire system. The data can be handled by using firebase real-time database. This data repository can be used further by applying machine learning algorithms as a future work [27, 28].

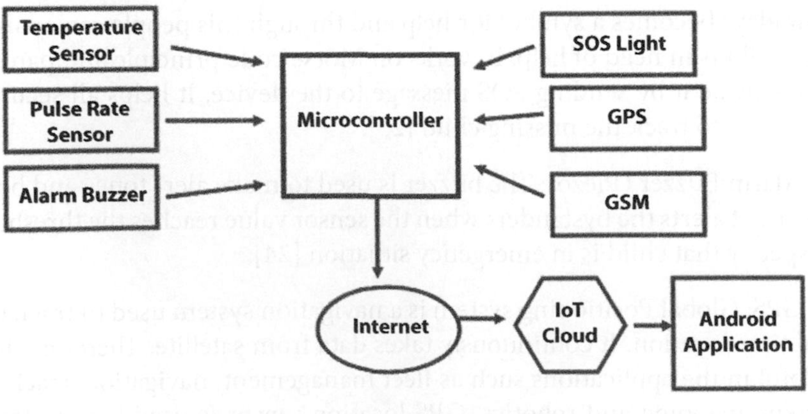

Figure 9.1 Conceptual architectural design of proposed system.

The motive of this proposed architecture to ensure the child's safety from violence. Therefore, communication of messages to required destination plays an important role.

The generalized algorithm for proposed model is shown in Figure 9.2.

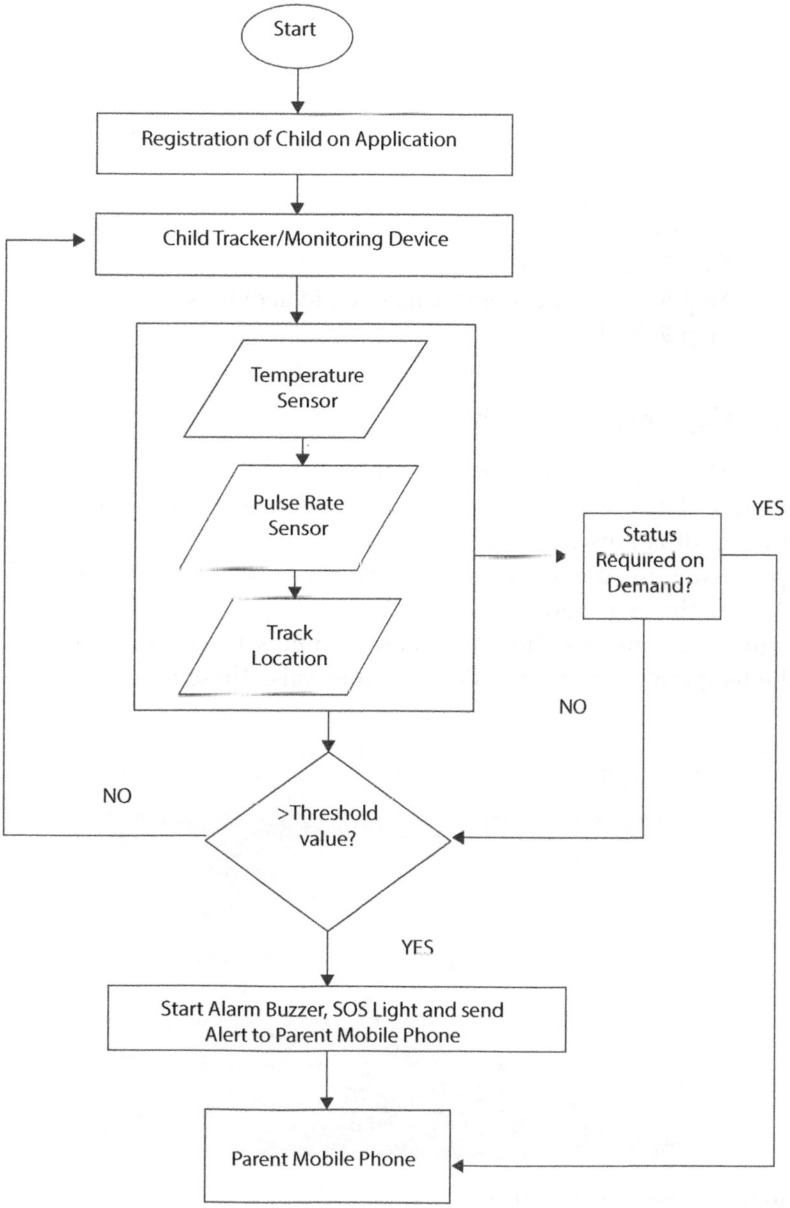

Figure 9.2 Demonstrates the necessary steps considered to track the child.

Algorithm:

Step 1: Registration of child information on the android application.
Step 2: After completion of the registration, track the child's location and other parameters through the child's monitoring device.
Step 3: If (sensor_value > threshold_value)
Step 4: Then (Alarm_Buzzer=true) and (SOS_Light=true) and (Alert_status=true)
Step 5: Else If ((sensor_value) < (threshold_value OR Status= =true))
Step 7: Then go to step 2;
Step 8: Continue monitoring of child activities
Step 9: End

9.2.2 Implementation Phases

Phase 1: Data Collection through sensors
In phase I, data are collected through the sensors that monitor the environmental situations and send the readings to the cloud managed database for processing in subsequent phases. We have used bolt Wi-Fi module to implement the methodology.

Figures 9.3 and 9.4 show the sensory inputs taken and the readings of the temperature sensor at different intervals. These readings from the

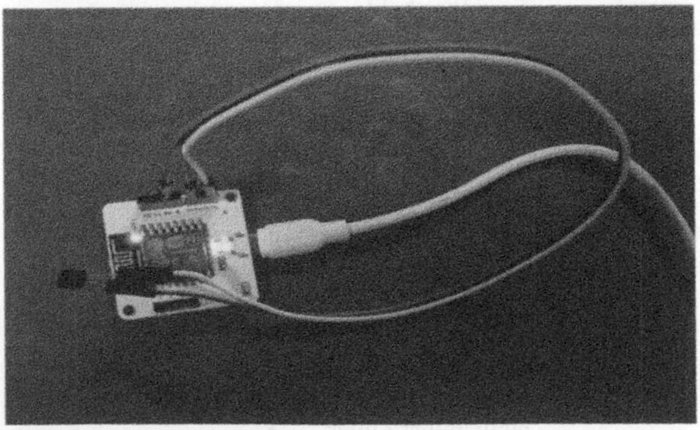

Figure 9.3 Sensory data collection.

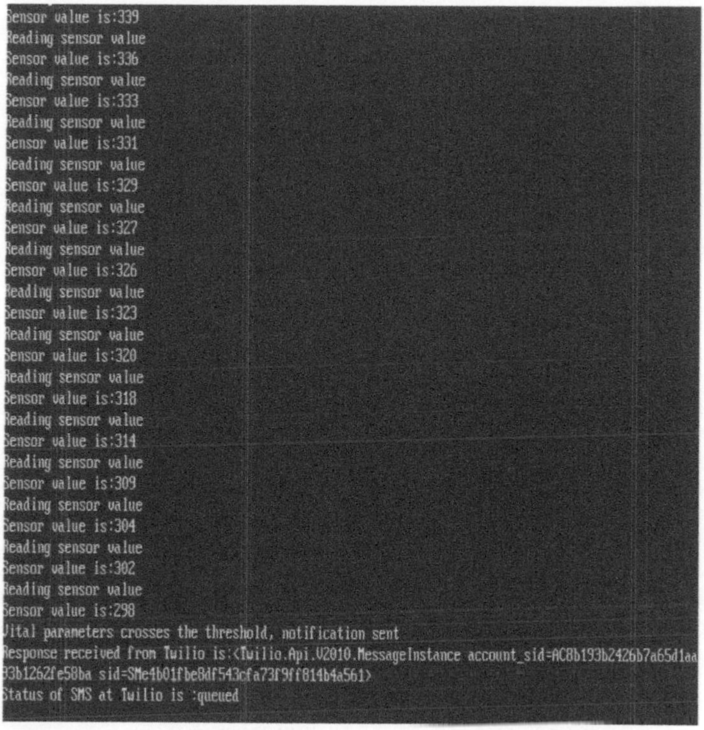

Figure 9.4 Readings from sensors.

sensors are then used further for comparison with the threshold value for initiating actions.

Phase II: Data storage on cloud
In the second phase of implementation, the data are stored on the cloud. All the sensory data collected in phase I are accumulated on to the Bolt

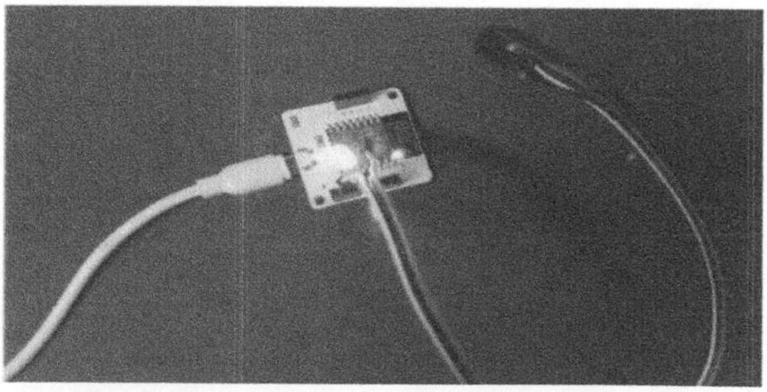

Figure 9.5 Turned ON buzzer.

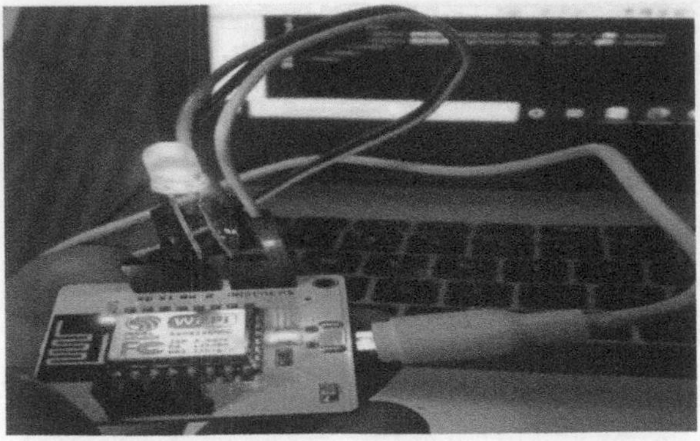

Figure 9.6 Turned ON LED.

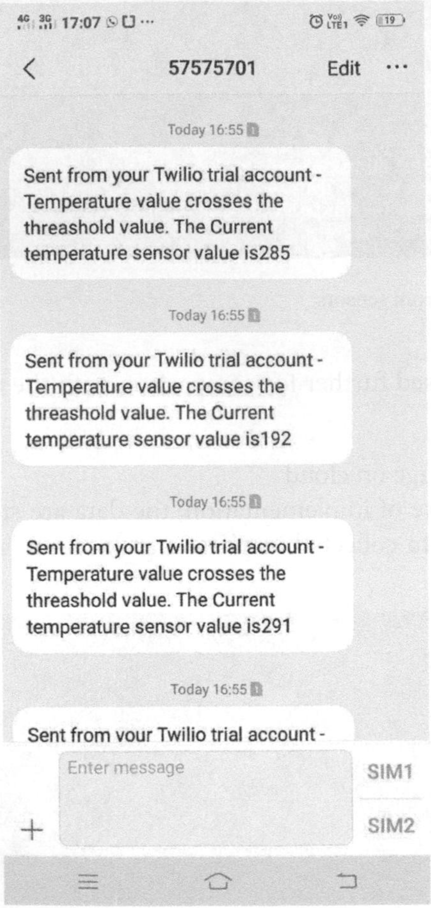

Figure 9.7 SMS notification via Twilio.

cloud for further process. It provides the flexibility for the enhancement in the future. We can also use it to apply machine learning algorithms to predict probable child activity readings in future.

Phase III: User Interface
It is the last phase where user the information on to their mobile phones. We use the trial account of twilio to provide sms notification to know the current readings and the child's vital parameters after goes beyond the threshold values. The subsequent figures show the results based on the readings taken by the sensors, which consists of turning ON the LED and Buzzer (Figures 9.5 and 9.6).

9.3 Advantages

The architectural design of the system provides various advantaged in various issues (Figure 9.7).

- It helps in providing the security against child abuse.
- It keeps real time tracking and monitoring of children and their well-being by measuring the vital parameters.
- It provides alert to the parents on their mobile phones in case of an emergency, and a child's exact location is detected with ease.
- It is a cost-effective smart system using Internet of things that provides total control and comfort.

9.4 Conclusion

The proposed work helps in providing solution for parents regarding safety of their child. It gives them a sense of security and able them to track all the activities of their child. The idea behind this proposal is to have a quick response in any of the circumstances without any delay. This particularly ensures to give better rescue to a child. This proposed idea focuses on wireless communication to give an alert and real time monitoring of individual [29, 30].

9.5 Future Scope

This chapter aims at real-time monitoring and tracking of child well-being. Therefore, real-time monitoring through visuals can make it easier to provide security. This can be facilitated by using web camera with the system. It can be used to identify and capture the offender and the identification can be sent to nearby police authorities for taking necessary action. Also, it can be used for tracking old people's safety and women's safety [31–33].

References

1. Shahid, N. and Aneja, S., Internet of things: Vision, application areas and research challenges, in: *2017 International Conference on I-SMAC (IoT in Social, Mobile, Analytics and Cloud), (I-SMAC)*, IEEE, pp. 583–587, February 2017.
2. Zeinab, K.A.M. and Elmustafa, S.A.A., Internet of things applications, challenges and related future technologies. *World Sci. News*, 2, 67, 126–148, 2017.
3. Bhattacharyya, R., Sociologies of India's missing children. *Asian Soc. Work Policy Rev.*, 11, 1, 90–101, 2017.
4. Coluccia, A. and Fascista, A., A review of advanced localization techniques for crowdsensing wireless sensor networks. *Sensors*, 19, 5, 988, 2019.
5. Moodbidri, A. and Shahnasser, H., Child safety wearable device, in: *2017 International Conference on Information Networking (ICOIN)*, IEEE, pp. 438–444, January 2017.
6. Kamalraj, R. and Sakthivel, M., A hybrid model on child security and activities monitoring system using iot, in: *2018 International Conference on Inventive Research in Computing Applications (ICIRCA)*, IEEE, pp. 996–999, July 2018.
7. Braam, K., Huang, T.C., Chen, C.H., Montgomery, E., Vo, S., Beausoleil, R., Wristband vital: A wearable multi- sensor microsystem for real-time assistance via low-power bluetooth link, in: *2015 IEEE 2nd World Forum on Internet of Things (WF-IoT)*, IEEE, pp. 87–91, December 2015.
8. Chowdhury, U., Chowdhury, P., Paul, S., Sen, A., Sarkar, P.P., Basak, S., Bhattacharya, A., Multi-sensor wearable for child safety, in: *2019 IEEE 10th Annual Ubiquitous Computing, Electronics & Mobile Communication Conference (UEMCON)*, IEEE, pp. 0968–0972, October 2019.
9. Sogi, N.R., Chatterjee, P., Nethra, U., Suma, V., SMARISA: A raspberry pi based smart ring for women safety using IoT, in: *2018 International Conference on Inventive Research in Computing Applications (ICIRCA)*, IEEE, pp. 451–454, 2018.

10. Gupta, P., Shah, D.D., Satyanarayana, K.V.V., An IoT framework for addressing parents concerns about safety of school going children. *Int. J. Electr. Comput. Eng.*, 6, 6, 3052, 2016.
11. Qureshi, A. and Anwaruddin, M., Child safety management using wireless technology. *J. Eng. Sci.*, 11, 1, Jan/2020. https://jespublication.com/upload/2020-110101.pdf.
12. Madhuri, M., Gill, A.Q., Khan, H.U., IoT-enabled smart child safety digital system architecture, in: *2020 IEEE 14th International Conference on Semantic Computing (ICSC)*, IEEE, pp. 166–169, February 2020.
13. Pramod, M., Bhaskar, C.V.U., Shikha, K., Iot wearable device for the safety and security of women and girl child. *IJMET*, 9, 83–88, 2018.
14. Bhanupriya, T. and Sundararajan, T.V.P., Activity tracker wrist band for children monitoring using IoT. *IJRITCC*, 5, 11, 52–57, 2017.
15. Ray, P.P., A survey on internet of things architectures. *J. King Saud Univ. Comput. Inf. Sci.*, 30, 3, 291–319, 2018.
16. Khan, J.Y. and Yuce, M.R. (Eds.), *Internet of Things (IoT): Systems and Applications*, CRC Press, Singapore, 2019.
17. Mell, P. and Grance, T., *Version 15 The NIST definition of cloud computing*, National Institute of Standards and Technology, Gaithersburg, MD, October 7 2009, http://csrc.nist.gov/groups/SNS/cloud-computing.
18. Stergiou, C., Psannis, K.E., Kim, B.G., Gupta, B., Secure integration of IoT and cloud computing. *Future Gener. Comp. Sy.*, 78, 964–975, 2018.
19. Kocakulak, M. and Butun, I., An overview of wireless sensor networks towards internet of things, in: *2017 IEEE 7th Annual Computing and Communication Workshop and Conference (CCWC)*, IEEE, pp. 1–6, January 2017.
20. Patnaik Patnaikuni, D.R., A comparative study of Arduino, Raspberry Pi and ESP8266 as IoT development board. *Int. J. Adv. Res. Comput. Sci.*, 8, 5, 2350–2352, 2017.
21. Kazi, S.S., Bajantri, G., Thite, T., Remote heart rate monitoring system using IoT. Techniques for sensing heartbeat using IoT. *IRJET*, 05, 04,2018.
22. Gay, W., DHT11 sensor, in: *Advanced Raspberry Pi*, pp. 399–418, Apress, Berkeley, CA, 2018.
23. Santhi, S., Udayakumar, E., Gowthaman, T., SOS emergency ad hoc wireless network, in: *Computational Intelligence and Sustainable Systems*, pp. 227–234, Springer, Cham, 2019.
24. Mohanaprakash, K. and Sekar, T.G., A smart alarm system for women's security. *IJEMR*, 8, 6, 89–92, 2018.
25. Singhal, M. and Shukla, A., Implementation of location based services in android using GPS and web services. *IJCSI*, 9, 1, 237, 2012.
26. Vaidya, V.D. and Vishwakarma, P., January 2018, A comparative analysis on smart home system to control, monitor and secure home, based on technologies like GSM, IoT, bluetooth and PIC microcontroller with zigbee

modulation. *2018 International Conference on Smart City and Emerging Technology (ICSCET)*, IEEE, pp. 1–4.
27. Samie, F., Bauer, L., Henkel, J., From cloud down to things: An overview of machine learning in internet of things. *IEEE Internet Things J.*, 6, 3, 4921–4934, 2019.
28. Ghosh, A., Application of chatbots in women & child safety. *Int. Res. J. Eng. Technol.*, 5, 12, 1601–1603, 2018.
29. Abd-Elminaam, D.S., Smart life saver system for alzheimer patients, down syndromes, and child missing using IoT. *Asian J. Appl. Sci.*, 6, 1, 21–37, 2018.
30. Chen, L.W., Chen, T.P., Weng, C.C., iBaby: A mobile children monitoring and finding system with stranger holding detection based on IoT technologies, in: *Proceedings of the ACM SIGCOMM 2019 Conference Posters and Demos*, pp. 66–68, August 2019.
31. Poonkuzhlai, P. and Aarthi, R., Child monitoring and safety system using wsn and iot technology. *Ann. Rom. Soc. Cell Biol.*, 25, 4, 10839–10847, 2021.
32. Rajalakshmi, S., Deborah, S.A., Soundarya, G., Varshitha, V., Sundhar, K.S., Safety device for children using IoT and deep learning techniques, in: *Advances in Smart System Technologies*, pp. 375–390, Springer, Singapore, 2021.
33. Revathi, K. and Manikandan, T., IoT based shrewd monitoring framework for children and women safety. *SPAST Abstracts*, 1, 01, 2021. https://spast.org/techrep/article/view/1679.

10

Water Content Prediction in Smart Agriculture of Rural India Using CNN and Transfer Learning Approach

Rohit Prasan Mandal[1]*, Deepanshu Dutta[2], Saranya Bhattacharjee[2] and Subhalaxmi Chakraborty[2]†

[1]*Department of Computer Science and Technology, University of Engineering & Management, Kolkata, West Bengal, India*
[2]*Department of Computer Science and Engineering, University of Engineering & Management, Kolkata, West Bengal, India*

Abstract

Soil moisture prediction is one of the emerging fields in the domain of smart agriculture. Ample moisture levels are of high importance to yields; thus, plants will not grow and develop with inadequate soil moisture, which influences soil temperature and heat capacity and also prevents soil from weathering. Accurate soil moisture detection can lead to proper growth of all kinds of trees and crops. In this chapter, we have implemented a deep learning and few transfer learning algorithms, such as convolutional neural network (CNN), VGGNet (VGG-16, VGG-19), Inception v3, ResNet50, Xception for prediction of soil moisture based on a very small custom dataset of 626 field images of rural India over a period of 10 days, both with and without the effect of data augmentation. We have also demonstrated how these pretrained algorithms save both time and resources compared to traditional deep learning models. Here, multiclass classification is considered based on three different classes, viz. "dry," "wet," "extremely wet." The algorithms CNN, VGG-16, VGG-19, Inception v3, ResNet50, Xception have individually showcased an accuracy score value of 0.73, 0.58, 0.68, 0.80, 0.56, and 0.70 respectively. Comparing the accuracy of each model Inception v3 has outperformed over the other results. The experimental results reflect that the proposed method using

**Corresponding author*: rohitmandal814566@gmail.com
†*Corresponding author*: subhalaxmi.chakraborty@uem.edu.in

Loveleen Gaur, Vernika Agarwal and Prasenjit Chatterjee (eds.) Decision Support Systems for Smart City Applications, (167–188) © 2023 Scrivener Publishing LLC

augmented dataset and transfer learning have drastically improved the performance of classifiers with respect to without augmentation conditions.

Keywords: Soil moisture, data augmentation, transfer learning, CNN, accuracy

10.1 Introduction

In India, the majority of the population is dependent on agriculture [1]. In order to meet future food demands for a rising global population while minimizing environmental impacts remains a challenge, for which precision irrigation [2] strategies will play a critical role. In evolving approaches to agricultural innovations and farming practices, our approach is more prominent and plausible. If the model is used on images taken by drones could save some time and efforts of farmers. It is clearly visible that we can monitor water quality using deep learning, mainly autoencoders and deep belief network [3]. We can quantify water quality by the help of machine learning and deep learning algorithms. There are some parameters used as data like temp, dissolved oxygen (DO) (% sat), pH, conductivity, biochemical oxygen demand (BOD), nitrates (NO_3), fecal and total coli forms (TC). Here we see that machine learning plays a great role in predicting water quality index (WQI) [4]. We can warn the user before water gets contaminated. It consists of the physical and chemical sensor to measure pH, turbidity, color, DO, conductivity, etc. to check the parameters. The data collected from the sensors and used as data to feed the neural network for the analysis [5]. Different plants have different water requirements, and when we want to put it into production, it always requires experienced farmers for individual crops. Our approach is quite simple and straightforward, the unsupervised machine learning works on the images captured by drone and helps cultivators work remotely using TensorFlow [6]. In Sharma *et al.* [7], the authors implemented an ensemble-based architecture for identification of nutritional deficiency in rice plants. In recent years, various researchers have worked on numerous approaches in dealing with soil moisture prediction [8–10].

In this work, we have chosen transfer learning [7] approach because in case of conventional deep learning methods, it is designed to work in isolation and can solve specific tasks, the model has to rebuild from scratch when feature-space distribution changes while in transfer learning we can overcome the isolated training. More specifically transfer learning uses previous knowledge to do another set of tasks, which gives higher accuracy without training it from scratch. In this whole process we have worked with convolutional neural network (CNN) [8, 18], VGGNet (VGG-16, VGG-19) [9], Inception v3 [10], ResNet50 [11], Xception algorithms [12].

For the rest of the article, we have discussed about the background of transfer learning. Then we have discussed how we have used data augmentation [13, 19] approach with transfer learning. It has been noticed that, with VGGNet, it is promising rather than the traditional approach of CNN. Finally, the result and conclusion have been represented in contrast to the traditional CNN method [14].

10.2 Proposed Method

In this section, we have described our dataset and proposal to perform the experiments in the mentioned steps (Figure 10.1): (I) We have created our own custom dataset of 626 individual field images of soil under three different classes "dry," "wet," "extremely wet." (II) To convert the raw data into an understandable format for the model we have operated prepossessing by converting the dimension of the picture from 4000*1844 to 200*200. (III) To improve the performance of our transfer learning models on a small dataset we have used data augmentation to form new and different examples. (IV) We have trained our Transfer Learning algorithms, *viz.* VGG16, VGG19, INCEPTION, XCEPTION, ResNet50 with and without Data Augmentation [15]. (V) We have used softmax activation function in the output layer for this multi class classification.

10.2.1 Corpus Creation

We have created our own labeled custom dataset for this experiment by capturing rural Indian field images in RGB format which contains three

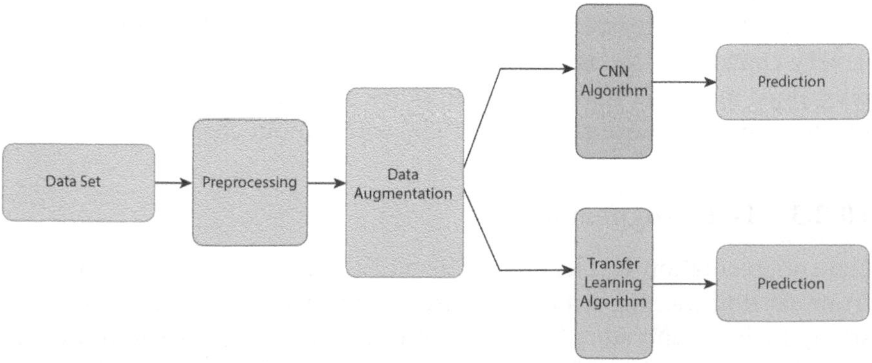

Figure 10.1 Proposed method.

Figure 10.2 Field images of wet, dry, and extremely wet test cases.

classes: *"dry," "wet," "extremely wet."* In the training set, the "dry" class contains 110 samples, "wet" contains 126 samples and "extremely wet" contains 74 samples. Figure 10.2 represents some data samples of different classes.

Figures 10.2, 10.2f, 10.2g are the examples of *Dry* condition. Figures 10.2a, 10.2b and 10.2h are examples of *wet* conditions, Figures 10.2c and 10.2e are examples of *very wet* conditions.

10.2.2 Data Pre-Processing

As pretrained transfer learning algorithms are trained on prepossessed sample images of ImageNet comes with the dimension of 256*256, we should resemble that dimension with our dataset samples accordingly. In our dataset, the dimension of most of the pictures are 4000*1844, which is very convoluted for any classifier to work on. So, while importing the pretrained model we have used *input shape* parameter to give the required image shape and convert it into 200*200 with 3 input channels.

10.2.3 Data Augmentation

Data augmentation is a technique to artificially create new training data from existing training data. Keras uses ImageDataGenerator() to quickly set up python generators that automatically turn image files into preprocessed tensors that can be fed directly into models during the training.

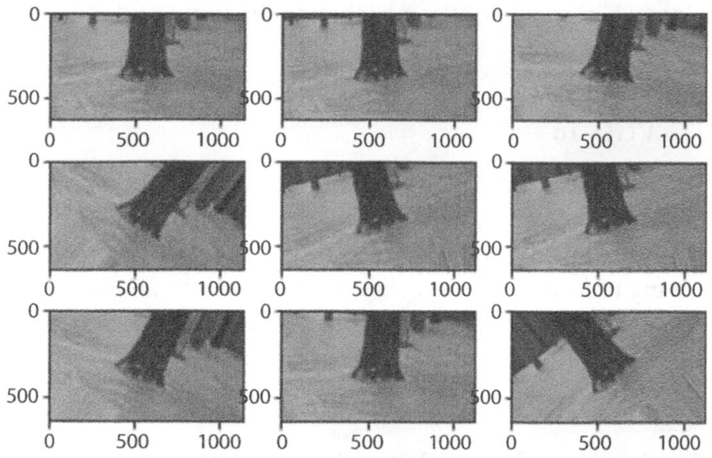

Figure 10.3 Artificially creating new training data.

Image transformation operations are applied on the training dataset, such as rotation, translation, and zooming to produce new versions of the existing images. In our study, the parameters are: rotation range = 40, height shift range = 0.2, width shift range = 0.2, shear range = 0.2, zoom range = 0.2, horizontal flip = True. From Figure 10.3, we can observe that this is more effective when we have very less amount of data. There are different commonly used activation functions like Relu, Sigmoid, Softmax, Tanh. For most of the cases of binary class classification sigmoid is used and for multiclass classification softmax is used.

10.2.4 CNN

We have chosen the traditional CNN approach in the first place, where the *Convolution layer* is extracting various important features from the input images. The second layer, *pooling layer, is reducing* the size of the feature map to reduce the computational cost. And the last layer, *fully connected layer*, where the output of the pooling layer acts as input to the fully connected layer.

10.2.5 Transfer Learning Algorithms

Studies show transfer learning algorithms are apparently faster than conventional methods because it was pretrained on millions of images of ImageNet dataset. So, when we train the model on our dataset, we use existing ImageNet weights. We only change the input and output layer for our training purposes.

10.2.5.1 VGG-16

VGG-16 is a convolutional neural network which has 16 layers. In Figure 10.4, the layers are *input layer, conv2D, maxpooling, flatten, dense,* and *output* layer with 1000 units. It is built as a deep CNN, VGG-16 outperforms baselines on many tasks and datasets outside of ImageNet, and in this research, we have used our own custom dataset where the results are promising.

10.2.5.2 VGG-19

VGG-19 is also a specific convolutional network designed for classification and localization. As shown in Figure 10.5, the architecture has 19 layers,

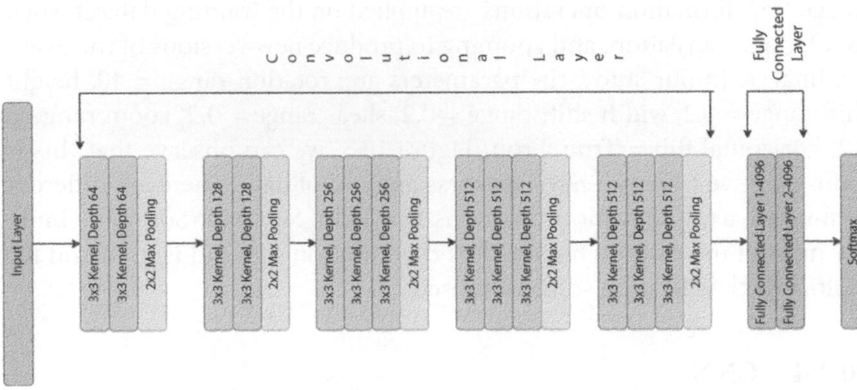

Figure 10.4 VGG-16 architecture.

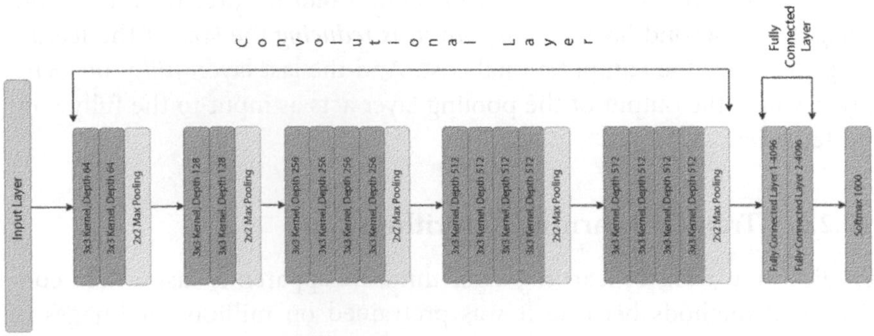

Figure 10.5 VGG-19 architecture.

Table 10.1 Comparative analysis in terms of accuracy with data augmentation.

	Loss	Accuracy	Validation loss	Validation accuracy
CNN	0.6584	0.6934	0.7106	0.7326
VGG-16	0.4611	0.8086	1.1704	0.5856
VGG-19	0.3754	0.8339	1.2354	0.6833
Inceptionv3	1.4113	0.8243	2.0019	0.8048
Resnet50	1.4783	0.5798	1.7653	0.5658
Xception	1.8149	0.8153	4.7580	0.7059

which means it has three more convolutional or weight layers. Although the pretrained network is able to classify images into 1000 object categories, we have classified field images into three classes. Results from Table 10.1 reflect that VGG-19 performed slightly better than VGG-16.

10.2.5.3 Inception v3

We have used Inception v3 because it uses several techniques for optimizing the network for better model adaptation and as a result it worked best on image classification. As shown in Figure 10.6, it has a total 42 layers.

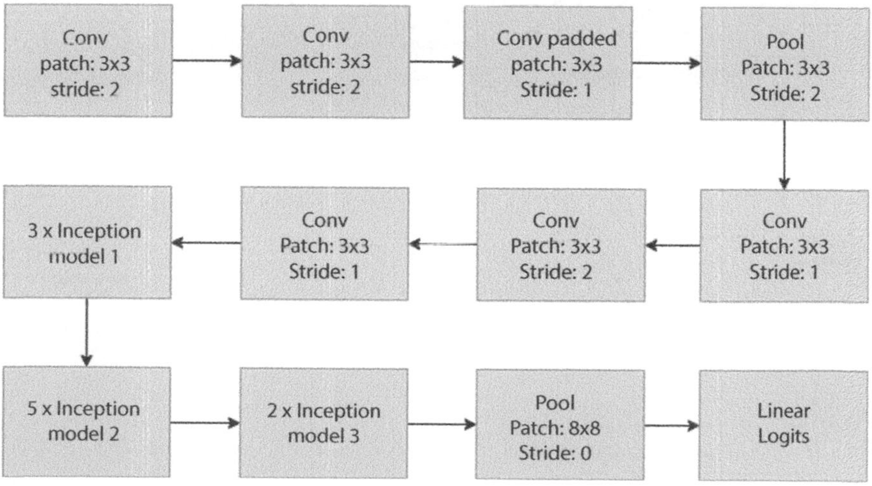

Figure 10.6 Inception model architecture.

The layers are conv, Conv padded, pooling, inception (modules 1, 2, 3), output layer with softmax activation function. In the case of our dataset, it outperformed all the other pretrained algorithms used in this chapter.

10.2.5.4 Xception

In xception model, there are two main parts shown in Figure 10.7, one is Depthwise Convolution and another is residual connection both of them really helps to improve the accuracy. Although in most of the classical classification cases it outperforms other transfer learning algorithms, in case of small datasets like ours, inception V3 resulted in maximum accuracy as shown in Table 10.1.

10.2.5.5 ResNet50

ResNet50 has two main parts, one is identity block and another is convolution block. Every convolution block has three convolution layers and each identity block also has three convolution layers. The ResNet50 has over 23 million trainable parameters. As we can observe from Figure 10.8, it has maximum number of layers than other models.

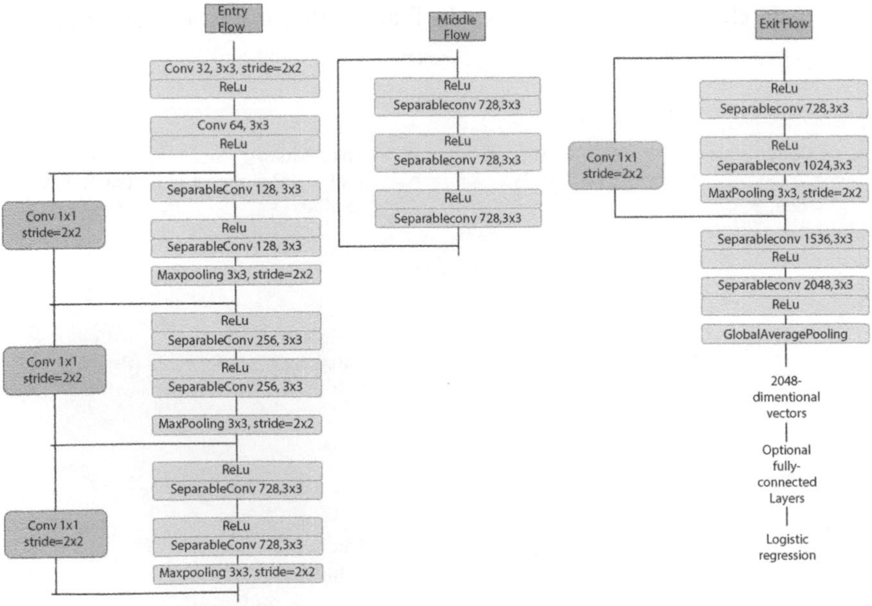

Figure 10.7 Architecture of xception model.

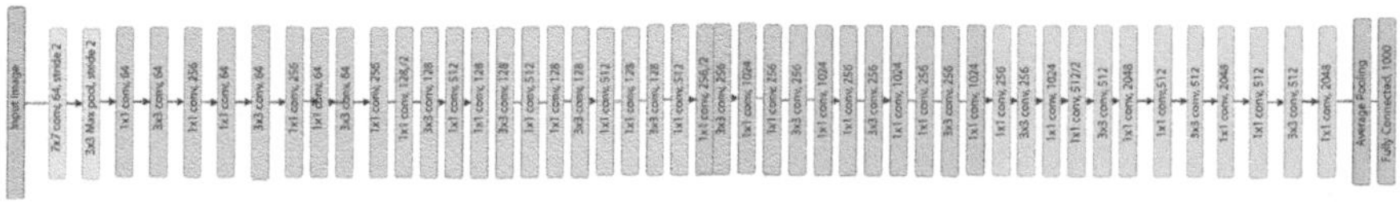

Figure 10.8 ResNet50 architecture.

10.3 Results and Discussion

In this section, we evaluated different performance metrics of various classifiers on original data and synthetic data obtained by applying augmentation methods. A description of the dataset is provided followed by the implementation of our experiment. Our simulations are implemented with TensorFlow 2.5.0 with Google Colab [16, 20] and Vega 8 2gb graphics, Ryzen 5-3500U, 8 GB RAM, Windows 10 Home 21H1. To be confident about the performance improvement of classifiers, we applied 6 different algorithms including Convolutional Neural Network (CNN) and Transfer Learning Algorithms: ResNet50, VGG-16, VGG-19, Xception, Inception v3. The classifiers are implemented using *tensorflow.keras.applications* package [17, 21].

10.3.1 Comparison of Classifiers

Here we have used 30 epochs to train our dataset. After data augmentation we have achieved these results. In Figure 10.9a, we can observe that the validation accuracy is increasing along with epochs and the overall validation accuracy is 73%, and the accuracy is consistent. In Figure 10.9b, we can conclude that the validation accuracy is increasing along with epochs and the overall validation accuracy is 80%, and it is also generalizing new test cases with maximum accuracy. In Figure 10.9c, the validation accuracy is haphazard along with epochs and the overall validation accuracy is approximately 57%. In Figure 10.9d, the validation accuracy first increases and decreases very much along with epochs and accuracy is not consistent, the overall validation accuracy is approximately 58%. In Figure 10.9e, validation accuracy increases along with epochs and not close to training accuracy and accuracy is 68%. In Figure 10.9f, validation accuracy increases then decreases and is not consistent along with epochs and accuracy is at about 70%.

Table 10.1 represents the accuracy score obtained over various classification algorithms for different sampling strategies, including imbalance data. The column *imbalance* represents the accuracy score of the data. As

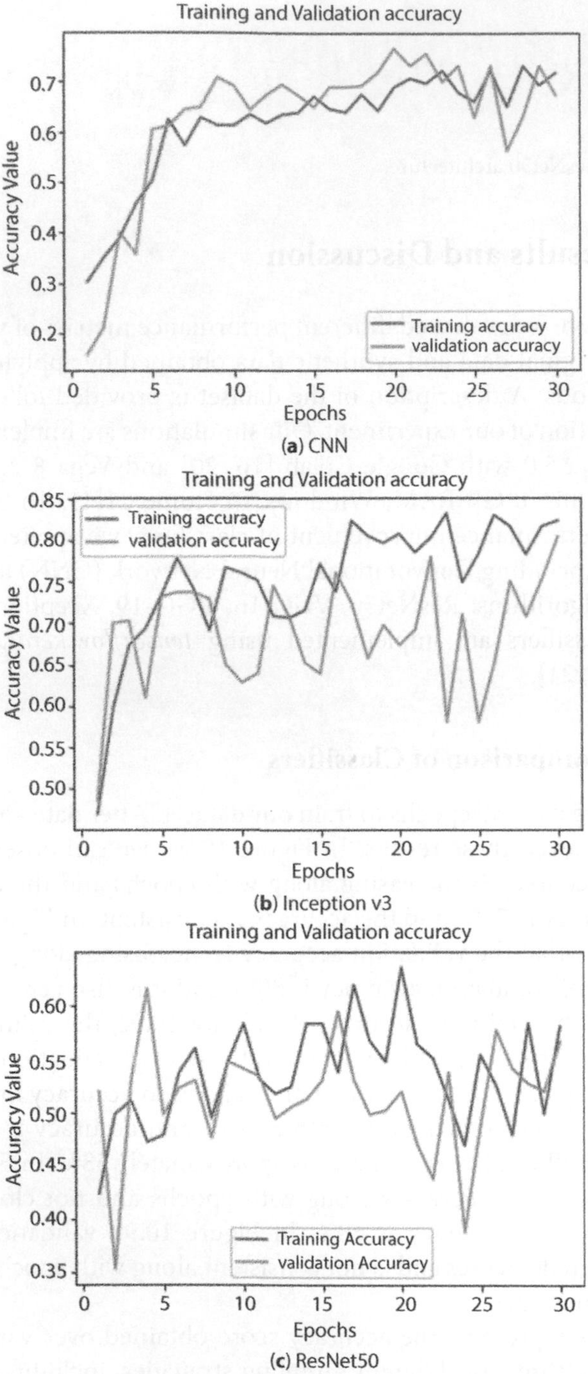

Figure 10.9 Comparison: validation accuracy. (*Continued*)

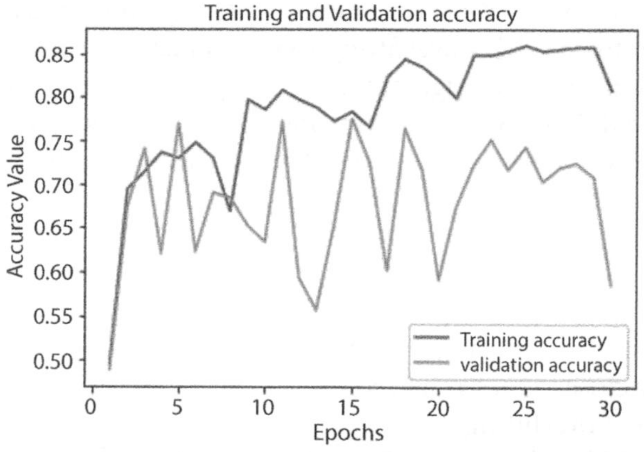

(**d**) VGG-16

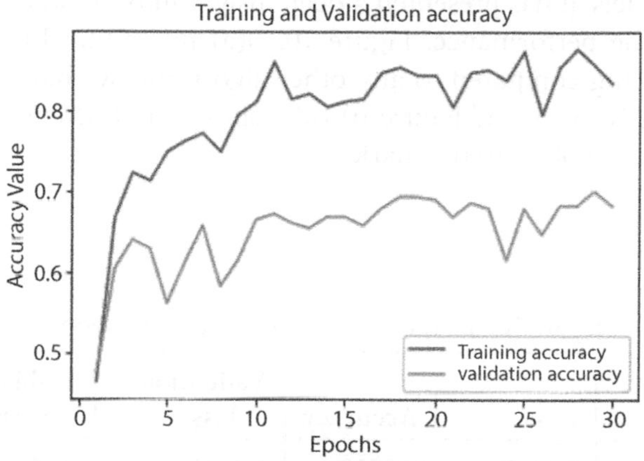

(**e**) VGG-19

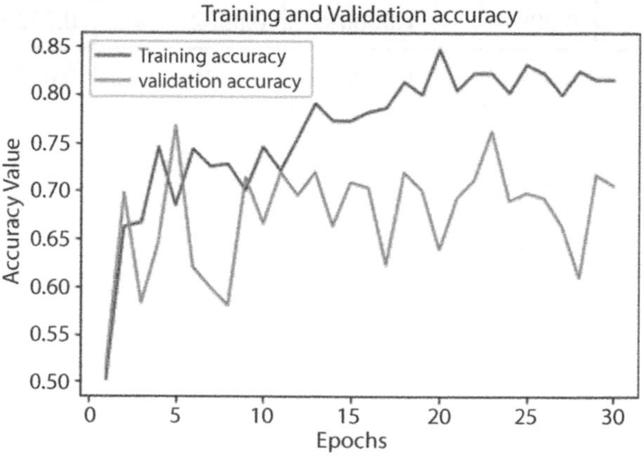

(**f**) Xception

Figure 10.9 (Continued) Comparison: validation accuracy.

shown, the accuracy score is very undesirable, they obtained almost 100% accuracy but the model is overfitting so the validation scores are very low. After applying the data augmentation, the performance of the data has been improved. It is noticeable from Table 10.1 that among all the classifiers Inception v3 performed best. All classifiers acquired an accuracy score of 0.73, 0.58, 0.68, 0.80, 0.56, and 0.70, respectively.

From Table 10.2, we can conclude that without data augmentation, we have observed even 100% accuracy in most of the algorithms but in each case validation accuracy explicitly indicates over-fitting. In Table 10.1, it is apparent that, after data augmentation, our actual accuracy and validation accuracy are close to each other, so we can conclude that we have overcome the problem of overfitting.

Figure 10.10 has documented an extensive review of the various transfer learning models, it has presented a deep insight into the study and helped us analyze the performance. Figure 10.10(a) infers that Inception v3 is more promising compared to any other algorithms we have used in our work. Figure 10.10(b) and Figure 10.10(c) show that ResNet50 exhibits low accuracy and is not up to the mark.

Table 10.2 Comparative analysis: without data augmentation.

	Loss	**Accuracy**	**Validation loss**	**Validation accuracy**
CNN	0.0327	0.9929	5.4647	0.6684
VGG-16	0.0097	1.0000	0.8582	0.7326
VGG-19	0.0097	1.0000	1.4446	0.6477
Incepsionv3	1.4624	1.0000	1.8793	0.7353
Resnet50	0.7241	0.8078	1.1483	0.6512
Xception	9.8642e-05	1.0000	3.2397	0.6417

Extreamly Wet 118.jpg

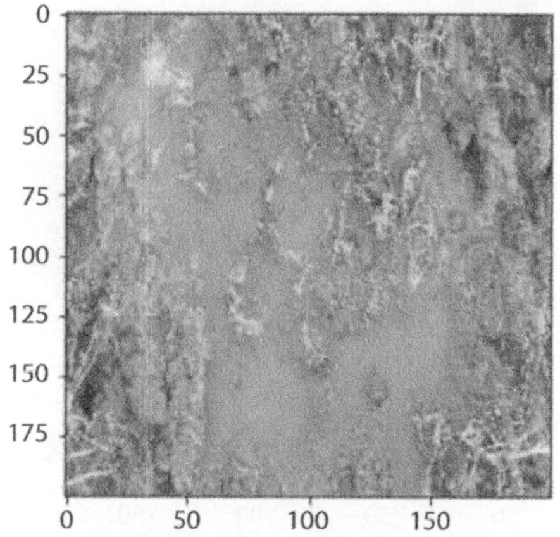

Extreamly Wet 119.jpg

(**a**) Inception v3

Figure 10.10 Prediction by different algorithms on test data with augmentation.

(*Continued*)

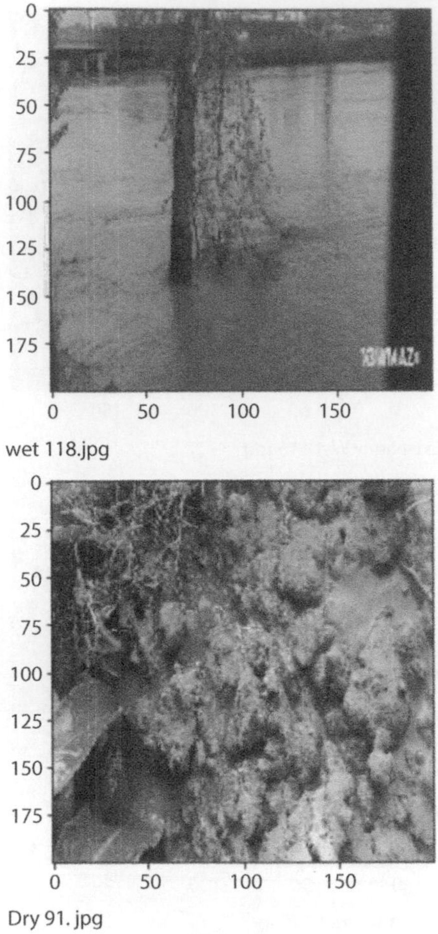

(**b**) ResNet50

Figure 10.10 (Continued) Prediction by different algorithms on test data with augmentation.

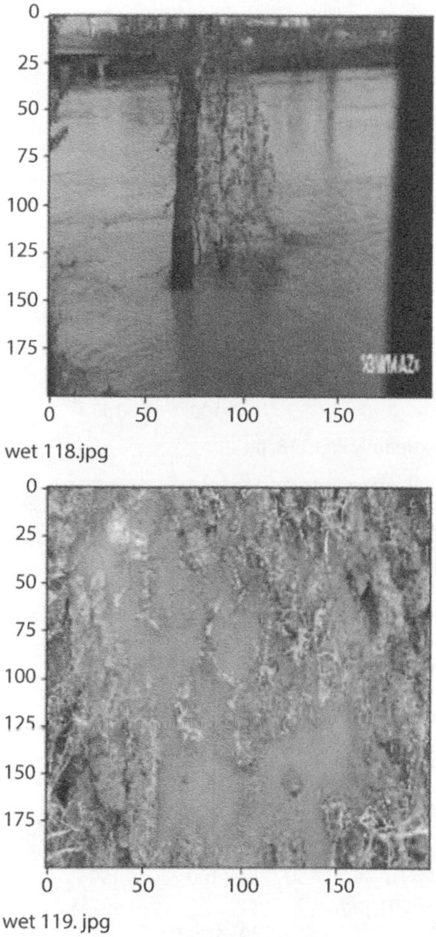

(c) VGG-19

Figure 10.10 (Continued) Prediction by different algorithms on test data with augmentation.

(**d**) Xception

Figure 10.10 (Continued) Prediction by different algorithms on test data with augmentation.

Extreamly Wet 118 . jpg

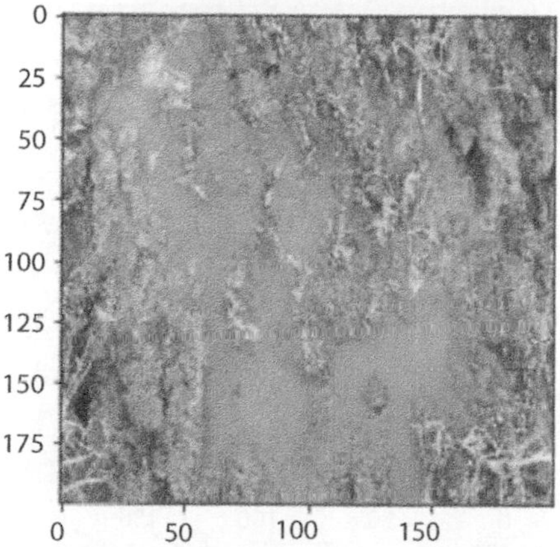

Extreamly Wet 119 . jpg

(**e**) CNN

Figure 10.10 (Continued) Prediction by different algorithms on test data with augmentation.

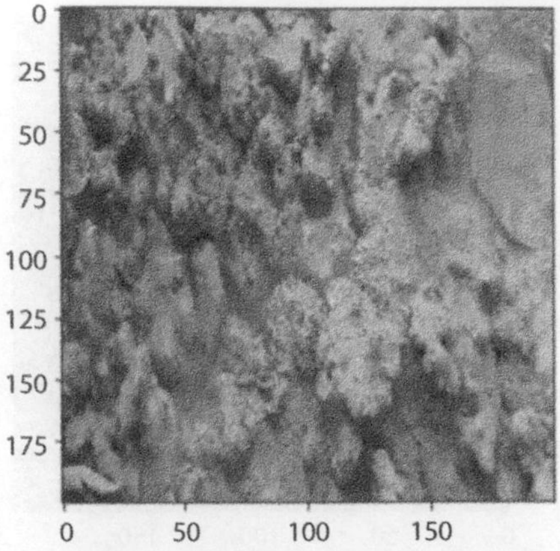

wet 100.jpg

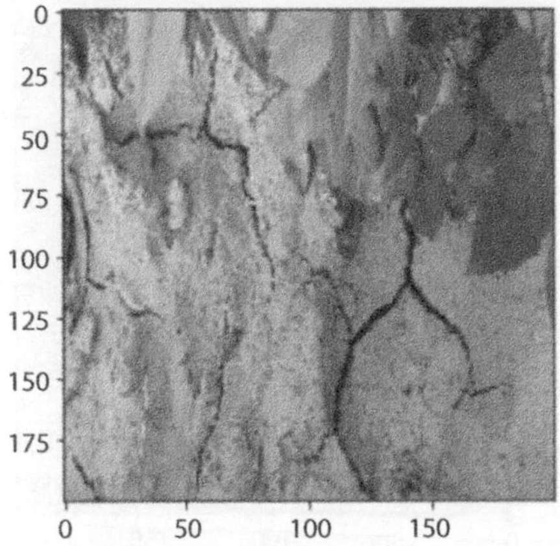

wet 121. jpg

(f) VGG-16.

Figure 10.10 (Continued) Prediction by different algorithms on test data with augmentation.

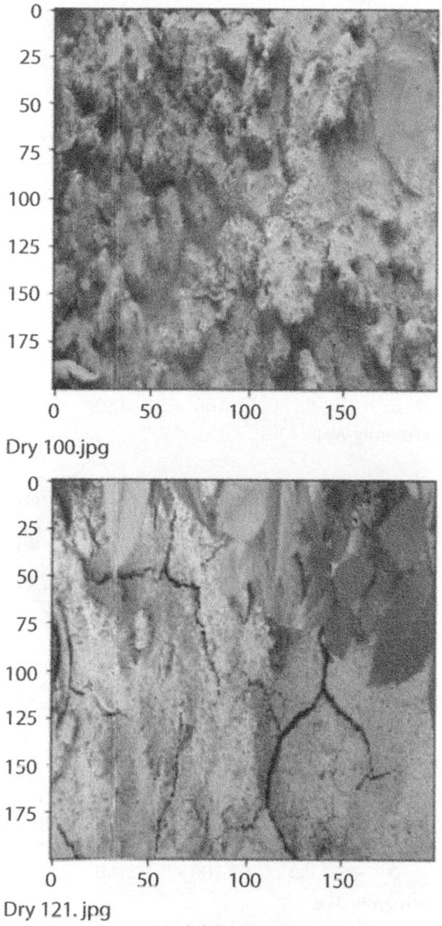

(g) ResNet50

Figure 10.10 (Continued) Prediction by different algorithms on test data with augmentation.

(**h**) Vgg-16.

Figure 10.10 (Continued) Prediction by different algorithms on test data with augmentation.

10.4 Conclusion

In this chapter, we have focused on the detection of water in agricultural land using CNN and five transfer learning algorithms. By using this custom dataset, all the six algorithms have been evaluated. The problem of overfitting is presently resolved by the utilization of data augmentation. The experimental results further show that combining an augmented dataset and transfer learning, the suggested strategy significantly enhanced the performance of classifiers. Another observation from our experiment is if our model works on images collected by drone could help agriculture on a large scale. Our research when incorporated with Internet of Things (IOT) could lead millions of farmers to have sustainable irrigation across India by minimizing the water waste and accurate prediction might gain an insight on how much water is required in a particular field. Our work could help different crops, depending on their requirements.

References

1. Suma, V., Internet-of-Things (IoT) based smart agriculture in India-an overview. *JISMAC*, 3, 01, 1–15, 2021.
2. Yao, Z., Lou, G., Zeng, X., Zhao, Q., Research and development precision irrigation control system in agricultural, in: *2010 International Conference on Computer and Communication Technologies in Agriculture Engineering*, vol. 3, pp. 117–120, IEEE, 2010.
3. Jodeh, S., Hamed, O., Melhem, A., Salghi, R., Jodeh, D., Azzaoui, K., Benmassaoud, Y., Murtada, K., Magnetic nanocellulose from olive industry solid waste for the effective removal of methylene blue from wastewater. *Environ. Sci. Pollut. Res.*, 25, 22, 22060–22074, 2018.
4. Ashwini, C., Singh, U.P. and Pawar, E., Shristi Water quality monitoring using machine learning and IoT. *Int. J. Sci. Technol. Res.*, 8, 1046–1048, 2019.
5. Khan, M.S.I., Islam, N., Uddin, J., Islam, S. and Nasir, M.K., Water quality prediction and classification based on principal component regression and gradient boosting classifier approach. *J. King Saud Univ.-Comput. Inf. Sci.*,, 2021.
6. Abadi, M., Agarwal, A., Barham, P., Brevdo, E., Chen, Z., Citro, C., Corrado, G.S., Davis, A., Dean, J., Devin, M. and Ghemawat, S., Tensorflow: Large-scale machine learning on heterogeneous distributed systems. arXiv preprint arXiv:1603.04467, 2016.
7. Sharma, M., Nath, K., Sharma, R.K., Kumar, C.J., Chaudhary, A., Ensemble averaging of transfer learning models for identification of nutritional deficiency in rice plant. *Electronics*, 11, 1, 148, 2022.
8. Chen, Y., Li, L., Whiting, M., Chen, F., Sun, Z., Song, K., Wang, Q., Convolutional neural network model for soil moisture prediction and its transferability analysis based on laboratory vis-nir spectral data. *Int. J. Appl. Earth Obs. Geoinf.*, 104, 102550, 2021.
9. Pinnington, E., Amezcua, J., Cooper, E., Dadson, S., Ellis, R., Peng, J., Robinson, E., Morrison, R., Osborne, S., Quaife, T., Improving soil moisture prediction of a high-resolution land surface model by parameterising pedotransfer functions through assimilation of smap satellite data. *Hydrol. Earth Syst. Sci.*, 25, 3, 1617–1641, 2021.

10. Fathololoumi, S., Vaezi, A.R., Alavipanah, S.K., Ghorbani, A., Saurette, D., Biswas, A., Effect of multi-temporal satellite images on soil moisture prediction using a digital soil mapping approach. *Geoderma*, 385, 114901, 2021.
11. Pan, S.J. and Yang, Q., A survey on transfer learning. *IEEE Trans. Knowl. Data Eng.*, 22, 10, 1345–1359, 2009.
12. Albawi, S., Mohammed, T.A., Al-Zawi, S., Understanding of a convolutional neural network, in: *International Conference on Engineering and Technology (ICET)*, IEEE, pp. 1–6, 2017.
13. Simonyan, K. and Zisserman, A., Very deep convolutional networks for large-scale image recognition, arXiv preprint arXiv:1409.1556, 2014.
14. Szegedy, C., Vanhoucke, V., Ioffe, S., Shlens, J., Wojna, Z., Rethinking the inception architecture for computer vision, in: *Proceedings of the IEEE Conference on Computer Vision and Pattern Recognition*, pp. 2818–2826, 2016.
15. He, K., Zhang, X., Ren, S., Sun, J., Deep residual learning for image recognition, in: *Proceedings of the IEEE Conference on Computer Vision and Pattern Recognition*, pp. 770–778, 2016.
16. Chollet, F., Xception: Deep learning with depthwise separable convolutions, in: *Proceedings of the IEEE Conference on Computer Vision and Pattern Recognition*, pp. 1251–1258, 2017.
17. Shorten, C. and Khoshgoftaar, T., A survey on image data augmentation for deep learning. *J. Big Bata*, 6, 1, 1–48, 2019.
18. Yamashita, R., Nishio, M., Do, R., Togashi, K., Convolutional neural networks: An overview and application in radiology. *Insights Imaging*, 9, 4, 611–629, 2018.
19. Mikolajczyk, A. and Grochowski, M., Data augmentation for improving deep learning in image classification problem, in: *2018 International Interdisciplinary PhD Workshop (IIPhDW)*, IEEE, pp. 117–122, 2018.
20. Bisong, E., *Building machine learning and deep learning models on Google cloud platform* (pp. 59–64). Apress, Berkeley, CA, 2019.
21. Chollet, F., others., Keras. GitHub. 2007: 90-95, 2015.

11

Cognitiveness of 5G Technology Toward Sustainable Development of Smart Cities

Kumari Priyanka, Gnapika Mallavaram, Archit Raj, Devasis Pradhan* and Rajeswari

Department of Electronics & Communication Engineering, Acharya Institute of Technology, Bangalore, Kartanaka, India

Abstract

The principle objective of this logical exploration is to analyze the role of information and communication technologies (ICTs) in the sustainable development of smart cities toward futuristic communication. Along these lines, we have involved expressive examination and basic near investigation to feature the effect of the new age of 5G advancements on the shrewd improvement of the fundamental regions that structure the design of a city. The main objective of this chapter is to present the advantages of incorporating 5G innovations for an overall smart, comprehensive, and long haul improvement of cities. However to economically profit from 5G technology, it is very fundamental that authorities of urban areas/states should make productive execution structures as far as possible, including mechanical framework, human resources, advancement, inside guidelines, and clients.

Keywords: 5G, smart cities, network capacity, energy efficiency, MIMO, QoE

11.1 Introduction

The main aim of the cities in global is to make smart and sustainable toward efficient communication with less utilization of energy. This aim has become basic in late a very long time because of the way that expanding metropolitan agglomerations have prompted over exploitation of assets, expanding contamination, and the requirement for public

**Corresponding author*: devasispradhan@acharya.ac.in

Loveleen Gaur, Vernika Agarwal and Prasenjit Chatterjee (eds.) Decision Support Systems for Smart City Applications, (189–204) © 2023 Scrivener Publishing LLC

administrations to meet the current social real factors. The ideal answer for reacting to the new difficulties is the constant combination of Information and Communication Technologies (ICTs) into the improvement of public administrations, so they can support a top-notch of life for residents. Consequently, the primary target of this logical article is to dissect how the ICTs add to the economical and long haul advancement of shrewd urban communities [1, 2].

In order to achieve this objective, we tried to attempt an observational survey in which we will attempt to decide the job of the 5G advances in supporting this turn of events and the necessities to be met by every city with the goal that these new advances produce the normal advantages. The development in urban area with sustainability this ideas create consciousness of the creation and utilization of assets needed for private, modern, transportation, business, or sporting cycles [26–29]. Maintainable metropolitan improvement confirms, focusing on ecological mindfulness in the utilization of regular assets in brilliant urban communities [30–32]. Yigitcanlar and Dizdaroglu [3] center around the idea of environmental urban areas in their examination. This idea has been created and advanced beginning around 1970 as a component of the maintainable improvement plan to make the city smart [3, 4].

11.2 Literature Review

The ideas of "5G," "smart cities" have been under center for the last decade or so for different reasons. They have been at the center of attention autonomous of one another; in any case, they have examined how they can possibly fortify and complete one another. The primary elements being presented by 5G are high network, lower inactivity, more prominent data transmission, and better help to portability. According to the reports, over 66% of associations will send 5G before the finish of 2020 [11]. According to the assessments of Cisco, before the finish of the year 2020, 5G is relied upon to create multiple times more traffic contrasted with 4G. 5G has been conveyed in different nations at a quick rate. According to the insights, practically 8% of LTE administrators have as of now sent off 5G administrations [12].

Likewise, brilliant urban areas is being acknowledged at a high speed as well. With the expansion of urbanization around the world, the idea of brilliant urban areas guarantees successful also effective methods for dealing with the always extending urban areas. In North America alone, the interest in brilliant urban areas related projects is relied upon to reach $244.5 by 2021 [13]. While 5G and smart city-related ideas are being carried out, the feelings of dread of expanded information traffic are stressing the scientists and financial backers the same. According to the white paper distributed by Cisco, the overall yearly development in the information traffic from 2017-2022 is relied upon to be 26%, and it will contact the sign of 122 EB each month by 2022 [10]. In spite of the fact that there has been a sharp expansion in the gadgets' stockpiling and calculation limits, portable gadgets cannot, in any case, cycle, store, and impart a gigantic measure of information needed for different applications.

11.3 5G: Overview

Worldwide energy-related carbon emission byproducts are on course to surge by 5% to 33 billion tons in 2021. Environmental change can possibly compromise each part of our regular routines, and organizations have an obligation to assist with making a more reasonable future. The rollout of the 5G network across the globe comes all at once where support ability ought to be at the bleeding edge of each basic business choice. It is currently an intense emergency. Environment and regular asset insurance are turning into a basic space of concern inside pretty much every honest organization's methodology [5, 6]. Figure 11.1 discusses the global ICTs connection from 2019 to 2030.

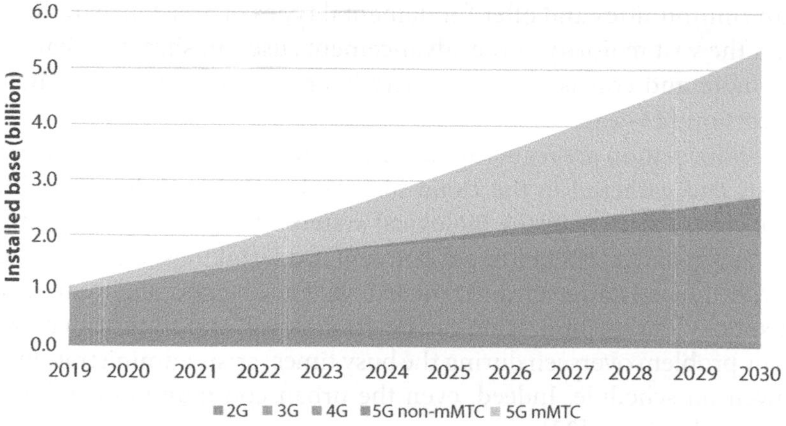

Figure 11.1 Global ICTs connection 2019-2030 (Source: Transforma Insights, 2020).

By 2050, the Earth's populace is relied upon to increment from 7 to 10 billion, and the world economy is projected to almost fourfold, spurring a developing interest for energy furthermore regular assets. This new economy is projected to utilize 80% more energy by 2050. As the total populace develops, we further strain the planet's assets, more products while adding to ecological waste, contamination, and decrease. 5G is more competent in the air interface. It has been planned with a drawn-out ability to empower cutting-edge client encounters, engage new organization models and convey new administrations. With high speed, prevalent dependability, and unimportant idleness, 5G will extend the portable biological system into new domains. 5G will affect each industry, making more secure transportation, distant medical services, accuracy farming, digitized coordinated operations with more realistic features [5–7].

11.4 Smart Cities

The thought of "smart cities" is in effect prevalently utilized nowadays from one side of the planet to the other. In this part, we clarify this idea exhaustively and with regards to its utilization for mankind. We can allude to the shrewd city as a "structure." This "system" is significantly framed with the assistance of state-of-the-workmanship innovations (data and correspondence). The inspiration driving this plan to be created and to be carried out to deal with the expanding pace of urbanization. It is getting moving for the legislatures to make due metropolitan urban communities and offer fundamental types of assistance to the residents. The vast majority of the advancements used in shrewd urban areas are remote and consistent and generally are conveying without human intercession [23–25].

The information are regularly communicated from different sensors and protests and gathered in the cloud for examination and direction. There are different partners in the biological system of brilliant urban communities, for instance, residents, government, undertakings, districts, and so on, with the assistance of different IoT gadgets, stages, and systems in a brilliant city, energy utilization can be diminished, and traffic can be without any problem overseen during the busy times, crisis administrations can be given on schedule. Indeed, even the urban communities can become cleaner and greener [23].

11.5 Cognitiveness of 5G Network

5G will do considerably more than altogether further develop a sustainable network for faster communication without any efficacy [8, 9]. It gives new freedoms, empowering us to convey noteworthy arrangements that span across society. Figure 11.2 shows the three basic constraints to enable sustainability.

11.5.1 Advancement of Society

5G opens state-of-the-art methods of further developing security and manageability. Some of the features are listed below:

- More astute power matrices for enormously diminished fossil fuel byproducts
- More associated vehicles sharing information to forestall street crashes
- Quicker sending of crisis administrations to mishaps
- Associated sensors that can distinguish and caution of cataclysmic events early
- Drones turn into a vital apparatus to speed up and uphold in emergency situations.
- Smart healthcare through which patients can remotely access sensitive documents from the medical institutions.

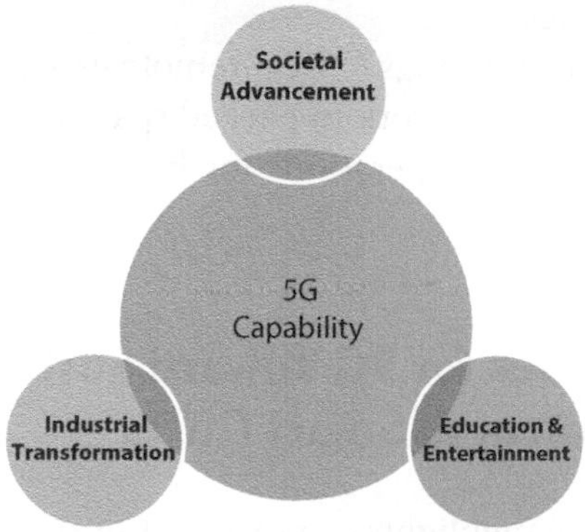

Figure 11.2 Basic constraints.

11.5.2 Industry Transformation

5G is the establishment for adaptable, proficient, and mindful business. With the help of the network following markets activity can be managed for the smart cities:

- Creation lines independently respond to the organic market
- Digitized caution about genuine hardware flaws early
- Strategic intelligent networks independently direct products dependent with respect to real-world situations.
- Full discernibility down to the singular thing at stockrooms and ports
- Remote admittance to strong robots and vehicles for further developed security in hazardous conditions
- Expanded utilization of IoT in horticulture to effectively develop crops

11.5.3 Amalgam of Education and Entertainment

5G makes way for more vivid diversion and seriously captivating training. Following are the emerging fields:

- More noteworthy authenticity in VR, AR, and expanded reality (XR) with lighter gadgets
- Conveying tangible encounters, similar to contact, through gadgets
- Additional drawing in strategies for instructing through vivid substance
- Vivid virtual gatherings to help remote group usefulness
- Steady and solid network in packed spaces
- New points and cooperations for live and far off occasion observers

11.6 Key Features for 5G Toward Sustainable Development of Smart Cities

5G, with its high-level highlights as portrayed in the previous segment, is very much situated to address the requests of the Smart City [9, 10]. The 5G elements that empower IoT reception on an enormous scope, needed for Smart Cities are:

a) **Connectivity between Devices:** Serving a tremendous expansion in the number of gadgets like sensors, cameras, actuators, and so forth associated with remote organizations. These are needed in homes, streets, traffic intersections, public spots like transport stands, rail route stations, air terminals, and so forth. This will uphold smart traffic systems, smart homes, public healthcare, security, and surveillance necessities of smart cities [9].
b) **Provide Large Bandwidth:** Extremely enormous transfer speed is needed to help both uplinks and download video-rich administrations over remote organizations and to help high information volumes. This is upheld by 5G.
c) **Latency:** Ultra-Low latency is needed for improved client encounters possibly including the conveyance of 3D pictures and 3D images and for applications like driverless vehicles. The idleness of under 1mS given by 5G backings this necessity [8, 9].
d) **Network Grid Enabling** - the 5G network can be divided into subnetworks that can be adjusted to support unique apps such as autonomous vehicles (AV), intelligent metering, public Wi-Fi, the appropriate allocation of utility resources, and so on [10].
v) **Energy Efficiency:** The gigantic expansion in associated gadgets making up a full-grown IoT is probably going to require preferred energy efficiencies over right now conceivable, for certain portable broadband gadgets needed to be on constantly while others will turn on irregularly [10, 11].

11.7 Application Enabled by 5G

5G advancements are relied upon to incorporate an assortment of gadgets, foundations, structures, items, and administrations and along these lines, to give ideal and manageable answers for addressing metropolitan issues like high energy utilization, high CO2 emanations, traffic, wellbeing framework, etc. As it can uphold the investigation of a lot of information, 5G technology can uphold complex systems and ventures through which metropolitan settlements can transform into intelligent urban communities, that guarantee the advancement of expectations for everyday comforts for metropolitan residents [13–15].

Among the arrangements that 5G advances ca, we notice public stopping frameworks and shrewd vehicle, lessening energy utilization and related expenses by utilizing savvy public lighting ideas, legitimate water and waste administration, progressed portable applications through which the gadgets and independent robots can be associated, the improvement of new programming and modern innovations, etc. Likewise, 5G advances will give the chance to foster new open administrations that right now cannot be enough conveyed (eg telehealth), and will have high strength on crucial occasions (eg, cataclysmic events) [15–17] (Table 11.1).

Table 11.1 Discuss the application of 5G communication in the sustainable development of smart cities.

SL. no	Identified fields	Benefits
1	Healthcare	General clinical assessments by means of cell phones; – Lessening hospitalization rates by remotely observing the advancement of patients' health status; – More extensive reception of computerized gadgets for healthcare observation.
2	Transportation	Diminishing time spent in rush hour gridlock; – Effectively distinguish free parking spots; – Diminishing metropolitan agglomeration levels; – Fast correspondence between vehicles to report traffic issues.
3	Security	Through a high-data transmission, quick and dependable correspondence can be carried out with the goal that offenses will be stayed away from or settled in an opportune manner; – Ultra HD video comments will permit specialists to utilize facial acknowledgment to recognize wrongdoers; – Cautioning frameworks created dependent on 5G sensors will give continuous data to individuals about setting off a catastrophic event like fires or floods

(Continued)

Table 11.1 Discuss the application of 5G communication in the sustainable development of smart cities. (*Continued*)

SL. no	Identified fields	Benefits
4	Retail Operation	Diminishing time and cost-related with shopping, by buying all items and administrations on the web; - Retailers will better oversee stores of items and build up a value methodology by gathering market request information at a time.
5	Environment	Exactness checking the degree of air and water contamination by introducing high accuracy sensors at central issues of the city; - Expanding the specific reusing rate; - Embracing manageable systems against environmental change
6	Energy	Building up quick associations between gadgets to oversee energy utilization productively, and subsequently decrease related expenses; - At the point when a voltage drop happens, there will be a continuous determination of the causes and prompt arrangements will be applied, for example, changing the heap in another gadget; - Decrease energy utilization in broad daylight space and related expenses by lessening light power when there is no traffic

11.8 Sustainable 5G-Green Network

This specific contextual analysis audits the most recent examination on green procedures for 5G organization and collecting energy for green correspondence, with dangerously request remote correspondence as referenced before, analysts searching for answers for getting together with the required prerequisite, because of that three ideas have arisen, these innovative ideas further develop throughput according to an alternate point of view which are:

a) Alleviate the TRX (transmitter-beneficiary) reach and lift recurrence reuse to machine correspondences by means of super thick organizations just as the machine.
b) Taking advantage of an inert and unregistered recurrence range in the unlicensed range through millimeter-wave correspondences just as long haul development.
c) Work on phantom productivity by utilizing a huge measure of antenna.

All the abovementioned new ideas instituted drain a lot of power, which would be vital in developing energy-efficient 5G networks, in this light low-energy consumption, may no longer be viable [18, 19].

11.9 Electricity Harvesting for Smart Cities

Accordingly, power reaping advancements, which license base stations and devices to collect energy from inexhaustible resources or even radio recurrence markers got sufficient consideration recently. Power reaping innovations give efficient power energy conveys replies to strolling assorted parts of remote correspondence networks. Thus, the significance of incorporating power reaping age in fate remote organization could not possibly be more significant [20, 21].

Basically, spectral efficiency is related with SINR as

$SE = K * B * N * \log2(1+ SINR)$....... (i)
where
K = reuse factor,
B = the signal bandwidth,
C = number of spatial beams,
D = Single link distance,
SINR = the signal to-interference-plus-noise ratio.

Also, energy efficiency is given as:
$EE = (SE)/(Pt + Pc)$..............................(ii)

Where
Pt = the consumed transmit powers
Pc = consumed circuit powers

In PTP links K, B, N, D are fixed, energy efficiency (EE) and spectral efficiency (SE) relation can be analyzed as:

a) Energy efficiency (EE) monotonically reduces with spectral efficiency (SE) If the energy consumption of the circuit is overlooked, that is when Pc is set to zero,
b) When Pc is greater than zero, the energy efficiency (EE) increases with spectral efficiency (SE) below a threshold and decreases with increasing spectral efficiency (SE) beyond the threshold.
c) As the spectral efficiency (SE) increases, the energy efficiency (EE) eventually congregates to the same values as for Pc = 0, because of the dominance of the transmit energy when no consideration is given to the circuit power.
d) Decreasing the circuit power will increase the energy efficiency (EE)-spectral efficiency (SE) trade-off region.

In realization, energy harvesting technologies can provide green energy, making it possible for 5G networks to perform at greater spectral efficiency compared to the normally restricted-energy networks [21–23].

11.10 Economic Impact of 5G Toward Sustainable Smart Cities

Smart city solutions applied to the administration of force matrices and vehicle traffic would bring about investment funds and advantages of many billions of dollars through a decrease in energy use, fuel utilization, and energy use. 5G arrangements would empower urban areas to decrease drive times, work on open well-being and create critical savvy lattice efficiencies [11]. 5G network would be constructed utilizing little cell organizations and would include 10 to 100 times more receiving antennas than 3G/4G organizations. Aside from conveying the fast and limits of 5G, these cells would uphold the expanded number of gadgets that would be associated with the organization of things to come [24, 25].

The new 5G network framework would itself make bunches of occupations. Legislatures should uphold the establishment of the new 5G foundation, as there will be a shift from the customary tall telecom pinnacles to little cell destinations introduced on light presents on utility costs. This might require an adjustment of the manner in which the current authorization process what is more charge structures. According to a review directed by the New Policy Institute, each change from the current age of portable correspondence to another age makes bunches of new positions open doors in its establishment and sending and furthermore from different administrations which are empowered from that age [12]. This would emphatically affect the GDP.

11.11 5G Challenges

High Data Rate (HDR): 5G targets incorporate raised information rates that are multiple times speedier than the greatest information rates possible. For the same purposes, precise information rates and potential are limited. Expansion in the use of a similar recurrence diminishes the absolute transfer speed accessible to a solitary gadget. Furthermore, the capacities of the spine, backhaul, and the total amount of data that can likewise be moved from base stations to cell phones, as well as the other way around, are restricted by front-pull existing in current gadgets. Measures will be needed to grow past current means rather than an additional productive use of the current means to influence information rate [12].
Quality of Experience (QoE): Basically, it is utilizing the network to his assumption for how the network ought to be. It is connected to raised information rate and low latency troubles, so it is aggregated in a specific way that makes a balance among them. High information rates and low latency can be an extraordinary streaming encounter, yet the energy utilization is not productive. In any case, a low QoE will drive to dissatisfactory assistance for the client [12].

11.12 Conclusion

The objective of the "coordinated vision" of smart cities is to further develop the citizens' quality of life in an economical way. It is critical that there is awesome coordination among different multipartners in the smart city eco-framework, to make it fruitful. Industry and administration gives cannot depend on the present 3G/4G remote frameworks for giving the objective vivid experience like reliability, short postponement, device energy productivity, and so forth, which are needed for smart city vision. 5G is fundamental for the execution of IoT that shapes the foundation of

Smart urban areas and thus turns into an empowering influence to the vision of smart cities.

5G will interface remote organizations to billions of gadgets, like vehicles, home apparatuses, hardware, and wearable innovation. Inventive regions will utilize smart city advances, like associated sensors, share more information to offer city types of assistance all the more proficiently and viably. Consequently, 5G will improve the "Internet of Things" and permit Smart Cities to create. The maximum capacity of Smart Cities can be opened by 5G organizations, making occupations and new organizations. 5G empowered smart cities can drive financial turn of events and further develop administrations and personal satisfaction for all networks in the urban areas.

References

1. Gartner, Gartner forecasts worldwide 5G network infrastructure revenue to reach $4.2 billion in 2020, *White paper from Gartner*, August 22, 2019. Available online: https://www.gartner.com/en/newsroom/press-releases/2019-08-22-gartner-forecasts-worldwide-5g-network-infrastructure (accessed on 17 August 2021).
2. International Telecommunication Union Radio Communication, *Detailed Specifications of the Terrestrial Radio Interfaces of International Mobile Telecommunications-2020 (IMT-2020); Recommendation ITU-R M.2150-0, 2021.02*, International Telecommunication Union Radio Communication, Geneva, Switzerland, 2021.
3. Yigitcanlar, T., Dur, F., Dizdaroglu, D. (2015). Towards prosperous sustainable cities: A multiscalar urban sustainability assessment approach. *Habitat International*, 45, 36-46.
4. UK Department for Digital, Culture, Media & Sport, *UK Telecoms Supply Chain Review Report*, Telecoms Security and Resilience Team Department for Digital, Culture, Media, and Sport, 100 Parliament Street SW1A 2BQ, July 2019.
5. Bansal, R., Obaid, A.J., Gupta, A., Singh, R., Pramanik, S., Impact of big data on digital transformation in 5G era. In *Journal of Physics: Conference Series*, 1963, 1, p. 012170. IOP Publishing, 2021 July.
6. Ahmad, I., Shahabuddin, S., Kumar, T., Okwuibe, J., Gurtov, A., Ylianttila, M., Security for 5G and beyond. *IEEE Commun. Surv. Tut.*, 21, 4, 3682–3722, 2019.
7. Huang, T., Yang, W., Wu, J., Ma, J., Zhang, X., Zhang, D., A survey on green 6G network: Architecture and technologies. *IEEE Access*, 7, 175758–175768, 2019.

8. Tomkos, I., Klonidis, D., Pikasis, E., Theodoridis, S., Toward the 6G network era: Opportunities and challenges. *IT Prof.*, 22, 1, 34–38, 2020.
9. Cui, J., Chen, J., Zhong, H., Zhang, J., Liu, L., Reliable and efficient content sharing for 5G- enabled vehicular networks. *IEEE Trans. Intell. Transp. Syst.*, 23, 2, Feb. 2022.
10. Loghin, D. *et al.*, The disruptions of 5G on data-driven technologies and applications. *IEEE Trans. Knowl. Data Eng.*, 32, 6, 1179–1198, June 1, 2020.
11. Agiwal, M., Roy, A., Saxena, N., Next generation 5G wireless networks: A comprehensive survey. *IEEE Commun. Surv. Tut.*, 18, 3, 1617–1655, 2019.
12. Lin, J.C.W., Srivastava, G., Zhang, Y., Djenouri, Y., Aloqaily, M., Privacy-preserving multi-objective sanitization model in 6G IoT environments. *IEEE Internet Things J.*, 8, 7, April 1, 2021.
13. Pradhan, D. and Priyanka, K.C., RF-energy harvesting (RF-EH) for sustainable ultra-dense green network (SUDGN) in 5G green communication. *Saudi J. Eng. Technol.*, 2020.
14. Pradhan, D., Sahu, P., Dash, A., Tun, H., Sustainability of 5G green network toward D2D communication with RF-energy techniques. *IEEE International Conference on Intelligent Technologies (CONIT 2021)*, K.L.E.I.T, Hubbali, Karnataka, 25-06-21, IEEE, IEEE Bangalore Section, pp. 1–10, 2021.
15. Pradhan, D. and Rajeswari, 5G-green wireless network for communication with efficient utilization of power and cognitiveness, in: *International Conference on Mobile Computing and Sustainable Informatics. ICMCSI 2020. EAI/Springer Innovations in Communication and Computing*, Springer, Cham, 2021, https://doi.org/10.1007/978-3-030-49795-8_32.
16. Tun, H.M., Lin, Z.T.T., Pradhan, D., Sahu, P.K., Slotted design of rectangular single/dual feed planar microstrip patch antenna for SISO and MIMO system. *Proc. of the International Conference on Electrical, Computer, and Energy Technologies (ICECET)*, Cape Town-South Africa, December 9-10 2021.
17. EESL (Energy Efficiency Services Limited) and IEA (International Energy Agency), *India's UJALA Story-Energy Efficient Prosperity*, EESL, New Delhi, 2017.
18. Hua, Y., Liang, P., Ma, Y., Cirik, A.C., Gao, Q., A method for broadband full-duplex MIMO radio. *IEEE Signal Process. Lett.*, 19, 12, 793–796, 2012.
19. Pradhan, D. and Priyanka, K.C., SDR network & network function virtualization for 5G Green communication (5G-GC), in: *Future Trends in 5G and 6G*, pp. 183–203, CRC Press, Boca Raton FL, 2022.
20. Devasis, P. and Priyanka, K.C., Effectiveness of spectrum sensing in cognitive radio toward 5G technology. *Saudi J. Eng. Technol.*, 4, 12, 473–785, Dec. 2019.
21. Cisco, Cisco visual networking index: Global mobile data traffic forecast update, *White paper from CISCO*, 2017-2022, Feb. 2019. Available At: https://www.Cisco.com/c/dam/m/en_us/network-intelligence/serviceprovider/digitalransformation/knowledge-networkwebinars/pdfs/1213-business-services-ckn.pdf.

22. Guevara, L. and Cheein, F.A., The role of 5G technologies: Challenges in smart cities and intelligent transportation systems. *MDPI J. Sustainability*, 12, 6469, 1–15, 2020.
23. Smya, S., Wang, H., Basar, A., 5G network simulation in smart cities using neural network algorithm. *J. Artif. Intell. Capsule Networks*, 3, 1, 43–52, 2021.
24. Ford, D.N. and Wolf, C.M., Smart cities with digital twin systems for disaster management. *J. Manage. Eng.*, 36, 4, 1–10, 2020.
25. Manimuthu, A., Dharshini, V., Zografopoulos, I., Konstantinou, C., Contactless technologies for smart cities: Big data, IoT, and cloud infrastructures. *SN Comput. Sci., A Springer Nat. J.*, 2, 334, 1–24, 2021.
26. Yigitcanlar, T., Sipe, N., Evans, R., & Pitot, M., A GIS-based land use and public transport accessibility indexing model. *Australian planner*, 44, 3, 30-37, 2007.
27. Pietrosemoli, L., & Monroy, C. R., The impact of sustainable construction and knowledge management on sustainability goals. A review of the Venezuelan renewable energy sector. *Renewable and Sustainable Energy Reviews*, 27, 683–691, 2013.
28. Goonetilleke, A., Yigitcanlar, T., Ayoko, G. A., & Egodawatta, P., *Sustainable urban water environment: Climate, pollution, and adaptation*. Cheltenham, UK: Edward Elgar, 2014.
29. Yigitcanlar, T., & Kamruzzaman, M., Investigating the interplay between transport, land use, and the environment: A literature review. *International journal of environmental science and technology*, 11, 8, 2121–2132, 2014.
30. Dizdaroglu, D., & Yigitcanlar, T., A parcel-scale assessment tool to measure sustainability through urban ecosystem components: The MUSIX model. Ecological Indicators, 41, 1, 115–130, 2014.
31. Yigitcanlar, T., & Teriman, S., Rethinking sustainable urban development: Towards an integrated planning and development process. International Journal of Environmental Science and Technology, 12, 1, 341–352, 2015.
32. Komninos, N., Smart environments and smart growth: Connecting innovation strategies and digital growth strategies. International Journal of Knowledge-Based Development, 7, 3, 240–263, 2016.

12

Society 5.0 and Authenticity: Looking to the Future

F.-E. Ouboutaib, A. Aitheda and S. Mekkaoui

National School of Commerce and Management, Research Team in Marketing Management and Territorial Communication, Agadir, Morocco

Abstract

Society 5.0 paradigm is at the core of recent research debates. Studies discuss the technological perspective and industrial practices. It embraces a huge need for a redesign to the traditional business into smart manners. It focuses on people, with the aid of digital technologies, who can involve a diverse lifestyle and aim for happiness in different means. The world became more and more digital. The changes are not in technological domain, human and society are changing too. However, the discussion about the consumer's point of view in this smart life is still emergent.

As commonly known, to realize a positive effect on key performance indicators, business approaches should not be limited to strictly technical aspects but should instead put the aim on determinants of consumer behavior. This research fills this gap. It designs a new concept in this framework in which the importance of knowledge is not determined exclusively by competitiveness and productivity, but by taking into account the consumer' demand. It uses a quantitative analysis in context of traditional products. The contributions aim to highlight synergies between technology, sociocultural, and economic systems.

Keywords: Authenticity quest, consumer, Society 5.0, Industry 4.0, Smartpls

**Corresponding author*: fouboutaib@gmail.com

Loveleen Gaur, Vernika Agarwal and Prasenjit Chatterjee (eds.) Decision Support Systems for Smart City Applications, (205–218)

12.1 Introduction

Technological advancements have been the driving force behind the advancement of human society. During the four industrial revolutions, the world has converted more and more digital. Changes are not occurring just in the technology field, the human and social fields are transforming too.

The fourth technological revolution, which materialize recently, is built on concepts and technology such as cyber-physical systems, the Internet of Things (IoT), and the Internet of services [1, 2]. Firms that take advantage of the opportunities and experiment with new technologies to see which ones improve operational efficiency and which ones improve consumer experience are more likely to succeed [3, 4]. For [5] this application, of new technologies and the transformation for processes, can enable the value creation, value capture and value offer for business models.

In the marketing field, the massive industrialization and uses of new technologies in the food sphere have generated new demands for consumers. Three main conditions, among others, are at the source of these which are: global health troubles, the birth of new consumer figures and the return to the embrace of products with a tribal flavor [6]. This global order has activated a quest for authenticity in different territories [6–10]. Likewise, the determinant of authenticity in online context represents a crucial topic on digital consumer behavior field [11], because studies emphasize the key role of technologies to sustainable growth and it can reinforce process innovation opportunities and management efficiency [12].

Closer to the focus of this chapter, these advanced technologies can transform the consumer experience by making the shopping more convenient and accessible, modifying how the customer shops, and/or influencing his interactions with digital tools. While industries continually process new technologies, the expectations of consumer come to hold concerning those technologies mutate [3]. In this sense, research underlines that the technological changes should be accompanied by significant changes in socioeconomic systems in order to avoid social cohesion weaknesses [13, 14]. Irrespective, research on this field is still in its infancy and confusing what key factors impact its potential use [12–14].

This fourth revolution industry is therefore both a technological and socio-economic phenomenon [14], firms have to comprehend how connected consumer products or services can operate as a critical basis for businesses to identify customer's determinants that influence their decision-making in regard to consuming connected products in the era of society 5.0. This practice leads to customer loyalty if trust is established between the customer and the organization [2]. Few authors have made empirical examinations of this phenomenon and its most important driving forces and barriers.

This chapter purposes to explore determinants of consumer digital quest of authenticity. This aim is important to accompany the digital transformation of the agricultural sector in Morocco [15]. For this purpose, it answers: to what extent is digital changing the consumer's quest for authenticity in the smart society era? It designs a quantitative study, with a sample of 300 Moroccan consumers. The analysis used a higher-order constructs modeling using the partial least squares structural equation modeling (PLS-SEM) in Smartpls3.

12.2 Theoretical Framework

12.2.1 A Brief History of Industry 4.0

Industry domain has endured over a time, a succession of "industrial revolutions" with cumulative complexity and productivity that reformed the existing practices. Over the past few years, industry 4.0 has arisen as an auspicious technology framework practiced for integrating and extending manufacturing approaches at both intra- and inter-organizational levels. It is very common that in its beginning the term "Industry 4.0" was publicly shared in 2011 in Germany. The group of representatives of business, political and scientific community was defined it as a very intelligent way to achieve the competitiveness of the industry through the strengthened integration of "cyber-physical systems" (CPS) into production's process [1]. Furthermore, Industry 4.0's emergence has been fuelled by the latest development in Information and Communications Technology (ICT). In this sense, the progress and the technological advances in 4th Intdustrial Revolution will offer a sustainable assortment of solutions to the huge needs of information in different manufacturing activities.

This period is marked by an important automation, digitization process and a huge use of information technologies (IT) that impact significantly changes in small and medium enterprises [2]. Several technologies have changed the dairy life such online stores and services that give opportunity to order directly from the store to the consumer's plate. With these technological advances, companies can enhance both operational efficiency and consumer experience and are likely to have successful business [3]. In this sense, this capability has been proven by the growing number of enterprises that have discovered the profits of digitizing horizontal and vertical chains and embraced Industry 4.0 in the process of mutating into leading digital enterprises in complex industrial ecosystems of the future [4]. It is very recognized that Industry 4.0 has a significant long-term strategic impact on global business development. In the global economy and business operations, experts have witnessed that there has been a need for this kind of practices to intensely increase the global level of industrialization, informatization, and manufacturing digitization to attain superior competency, efficiency, and competitiveness. In the study of Kagermann [5], industry 4.0 is determined as an umbrella term employed to draw the linking technological advances in the business environment. It is similar to the McKinsey's definition. It underlines that in this era the human-machine interaction is facing a new form of augmented reality and very fast transfer of information [16].

12.2.2 Marketing Authenticity in the Society 5.0 Era: Beyond Industry 4.0

Over time, the concerns of the research in the social field have advanced in parallel with the evolution of the industry domain. As we have known, these key inventions are the steam engine, the electricity, digitalization, and using information technology. They have concluded in the important progress produced in IT domain at the end of the 20th century with the genesis and mass generalization of the Internet which helmed to the fourth industrial revolution.

Researches in sociology are largely influenced by the different industrial movements. These different revolutions have marked the passage from modernity to postmodernity [6–18]. Certainly, there is no universal consensus for the definition of postmodernism, but researchers announce the birth of a very demanding consumer, changing and strongly influenced by new technologies [18]. This trend did not change the world, but it reinvented the way of seeing it. The considerable industrialization and uses of new technologies in the food sphere have caused an important quest of authenticity.

Technology advances have been accompanied many times with social problems. Researchers have underlined that 4th Industrial Revolution emphasizes is not a completely new phenomenon in human history; it shows some similarities with previous revolutions [13]. This implies necessity of deep social accompanied system which can remove the social problems that the world has known.

The main ideas of social orientation and technical innovations from Industry 4.0 were the basis for the concept of society 5.0. It is the answer to the demand of a new human-centered society in the fifth phase introduced by Japan [19, 20]. It looks beyond industry 4.0 to society 5.0. It can be characterized as a super-intelligent or creative society; it is following the hunting, agricultural, industrial, and information society [21]. It tries to transform beyond the industry and develop a super-intelligent society in which new knowledge and values are regularly designed to evolve the economic development and social welfare [22]. The present digital society is considerably more complex and interdependent than ever. For fifth society to be designed for the aims of sustainable development, it must rely on cooperation with stakeholders from various fields of knowledge, including economics, social science society and humanities [23]. It aims a world where basic goods and services are made accessible to everyone, anytime, anywhere, irrespective of region, age, gender, language, or other borders [24]. Society 5.0 mentions to employ advanced technologies and products to connect people and things, and to convey all kinds of knowledge and information in forming social chains, values, and innovative business in society [24, 25].

Keeping in mind that science and technology–based innovations have engendered some changes in society, but these transformations can be beneficial if society is well prepared for them. In this sense, industry should provide the true purpose which should contain social, environmental, and social considerations [21]. The society changes very rapidly to step in higher level like smart industry [13], but it deals with border time given to adjust with this very smart life [26, 27]. Therefore, managers are exposed to an issue: taking time for implementation these technologies or starting early and undertake fatal errors [28].

Figure 12.1 shows a transition step between from the only real social world to very digital social world. In this sense, we stress the importance of the interaction between real and digital, in this moment, to attain, for the future, important uses of digital or an only digital social life. In this

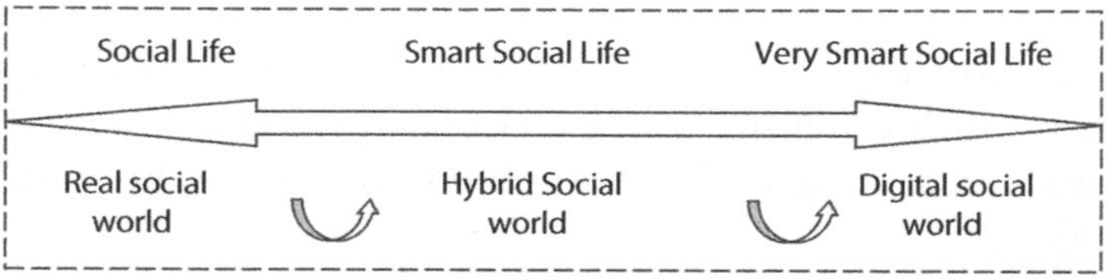

Figure 12.1 Transition to smart life.

way, its success is very linked to the technical feasibility and social system's adaptation [28, 29].

12.3 Research Design and Methodology

In socially connected world, research underlines that many customers seek social engagement and interactive experiences on platforms that foster a feeling of connection [30]. Furthermore, the perception of authenticity in digital context constitutes a crucial issue on consumer behavior [11–30]. Authenticity denotes the origin of the product and its producer [7]. It is a solution delivered to the postmodern consumer in order to attribute meaning to his food act [6, 9, 11] and appease his need for credibility and sincerity [31, 32].

The ability to link people and things, as well as the real and virtual worlds, will allow for the effective and efficient resolution of social challenges, as well as the enhancement of people's quality of life with healthy economic growth [33]. with the massive use of the web in the 4th industrial revolution [34], social presence does not require to be physically present for a person to experience its sense, but it can arise when products or situations trigger this feeling [35]. Literature suggests that in advanced retailer, consumer assumes a combined Omni-Channel experience with interaction between real and virtual world [35, 36]. As we noted, trust plays a key role in this context, we propose:

H.1. Authenticity's quest positively influences positive trust in real acquaintance
H.2. Authenticity's quest positively influences positive trust in digital acquaintance
H.3. Trust in real acquaintance positively influences positive trust in digital acquaintance
H.4. Trust in digital acquaintance influences positively intention
H.5. Trust in digital acquaintance positively influences positive real purchase intention

This research explores the influence of digital authenticity quest of consumer, which is defined as the consumer perception of origin, tradition, and unicity (Figure 12.2). It is mobilized as a higher-order construct [9].

Trust in digital acquaintance is defined as the trust in social media used by consumer. In Morocco, the number of social media users has exceed the

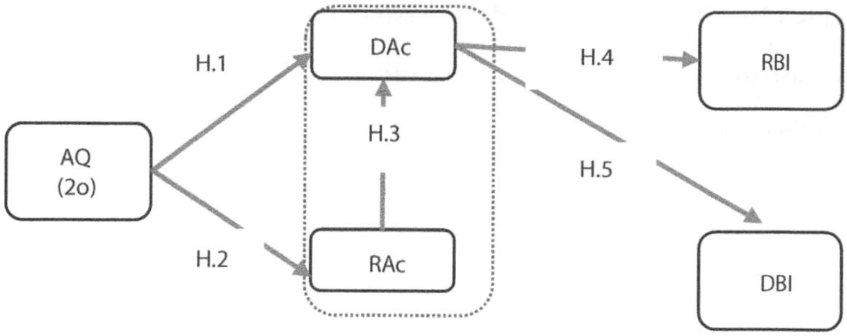

Figure 12.2 Research's model.

22 million users per day [37]. All variables' measure are taken from literature (see Appendix, Table 12.4).

The questionnaire of the study focused on the consumption of food products, specifically local products. This choice is motivated by two reasons: the importance of the origin in the definition of authenticity sought by the consumer and the importance of the local product in the digital transformation of the Moroccan agricultural sector. It was distributed to 300 consumers, convenience sample. This study used the partial-least-squares (PLS) path modeling approach with a higher-order construct [38, 39].

12.4 Results

12.4.1 Measurement Model

We have used the partial least squares structural equation modeling (PLS-SEM) in Smartpls3 to assess the measurement model. Assessment of measurement model involves examining the indicators loadings, reliability, convergent validity, and discriminant validity [38, 39]. We noted that in the first step, we have analyzed the measurement model with the lows factors of authenticity by using the two-stage approach of higher modeling [39].

Cronbach alpha and composite reliability are used to assess the internal consistency of all the items. In addition, the average variance extracted (AVE) and the composite reliability (CR) assessed the convergent and discriminant validity of the model. The respective results are displayed in Table 12.1.

Table 12.1 Measurement model assessment.

Construct	Item	Loading	Alpha	CR	AVE
Authenticity	OR HR SB	0.899 0.940 0.737	0.825	0.897	0.746
Trust in Real acquaintances	RAc1 RAc 2 RAc 3	0.936 0.959 0.943	0.942	0.963	0.896
Trust in Digital acquaintances	DAc1 DAc2 DAc3	0.905 0.899 0.842	0.862	0.914	0.779
Digital Behavioral intention	DBA1 DBA2 DBA3	0.802 0.828 0.803	0.740	0.852	0.737
Real Behavioral intention	RBA1 RBA2 RBA3	0.736 0.914 0.903	0.815	0.890	

Factor loadings transcend the recommended value of 0.7. AVE of all constructs varies between 0.737 and 0.896, which is above the recommended value of 0.5; and CR values surpass the threshold value of 0.7 in all cases.

Discriminant validity is the degree to which the constructs are different from each other, we have used the approach of Fornell and Larcker [38]. They suggested adequate discriminant validity when the square root of the AVE is larger than the corresponding correlations, which is satisfied for

Table 12.2 Discriminant validity.

	AQ	RAc	BAc	DBI	RBI
AQ	0.863				
RAc	0.601	0.946			
DAc	0.411	0.363	0.883		
DBI	0.546	0.475	0.484	0.811	
RBI	0.616	0.601	0.479	0.681	0.855

each construct showing adequate discriminant validity according to this criterion (Table 12.2 shows results of Smartpls).

Structural Model

To test the hypothesized relationships between the constructs, we obtain path coefficients, corresponding t values and p values by the bootstrapping procedure with a resample of 5000 in Smartpls (Figure 12.3). Table 12.3 shows the structural model coefficients of our conceptual framework. The general adjustment of the model met the standard criteria [38, 39]. An examination of p values in Table 12.3, suggesting the hypothesized relationships between constructs are statistically significant at the 1% level.

The first hypothesis regarding consumer quest of authenticity and trust in real acquaintances was accepted ($p<0.01$).

Real acquaintances influence positively the trust in digital consumer's acquaintances, but we note a weak effect (0.182). Furthermore, the important impact of authenticity in trust on digital denotes the importance of the first real experience for consumer. Researchers have pointed out that the use of digital technology differs according to the level of innovation of the company's sector [14].

The agricultural sector still lags behind other sectors, this is arising especially when regarding the small- and medium scaled firms and/or food processors that are linked with human skill, know-how, and technological constraints [40]. Results show that digital positively influences the behavioral intention of the consumer. We notice that the digital intention (0.484) is a little more important than the real intention (0.479). This result

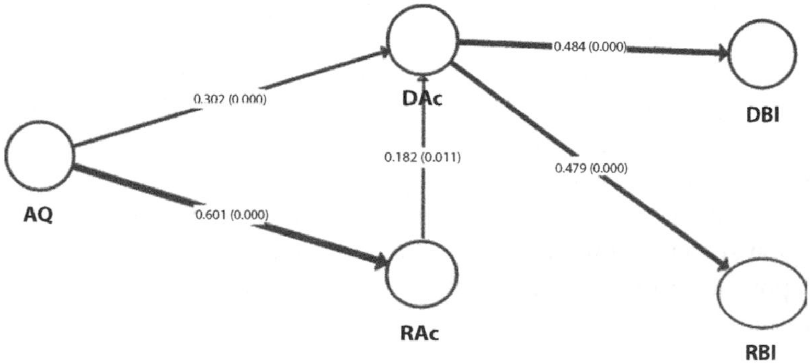

Figure 12.3 Structural model.

Table 12.3 Results of hypothesis testing.

Hypothesis	Path	T value*	Pvalue**	Confidence intervals	Decision
AQ -> RAc	0.601	13.473	0.000	0.520- 0.667	Accepted
AQ-> DAc	0.302	4.069	0.000	0.178- 0.421	Accepted
RAc -> DAc	0.182	2.358	0.009	0.060- 0.310	Accepted
DAc -> DBI	0.484	9.834	0.000	0.394 -0.558	Accepted
DAc -> RBI	0.479	10.494	0.000	0.394- 0.547	Accepted

*>1.96; **<0.01 [34].

emphasizes that we are still in a transition step (Figure 12.1) with a hybrid use of real and digital by the consumer [35, 36].

12.5 Conclusion

This chapter presents some of the application areas of industry 4.0 technologies which are pertinent to the food sector. Without a doubt, postmodern civilization has entered a new period of technical progress as a result of significant advances in mechanization, electricity, information, and networking technology [9, 12, 31, 32]. The ability to link people and things, as well as the real and virtual worlds, will allow for the effective and efficient resolution of social challenges, as well as the enhancement of people's quality of life with a healthy economic growth.

The new technologies' wave has already paved its method to food domain [40]. Some of the benefits are the durable contact and dialogue with consumer, cost reduction, market access, consumer satisfaction, and transparency. Postmodern society is undergoing a deep and rapid change [19–22], in line with researchers, we stress that without having enough time to adapt and establish solid bases to accompany this movement of technological development, especially for the food sector where the traditions and cultural system are very important.

The main of society 5.0 is the transition from digital manufacturing to digital society. However, many managers are facing issues in its effective application due to the rarity of a practical innovation management background.

Social innovation is an element of a regional innovation system, the meaning of knowledge is not determined exclusively by effectiveness and productivity, but by taking into account the consumers' demand. It draws on a real dialogue and cocreation of knowledge as part of consumer-producer partnerships.

By addressing social innovation practices from this dialogue, it fills the gap of the rarity of practical innovation that decreases the ability of producer. Social innovation engages the all ecosystem with a profound lecture.

The purpose of such technologies is the consumer-oriented adaptation of products and services that will increase value added for firms and customers in the same time. In this sense, we underline that people still need the real contact at least to accompany the change by giving the necessary time to the different stakeholders of the ecosystem. Business approaches should not be limited to strictly technical aspects but should instead put the aim on determinants of consumer behavior.

Though a key fact that needs to be seized is that the uses of the technologies will require the understanding of the product's difficulties, capabilities, and consumer's needs. When the aim of the technology meets the firm's need and fits within the consumer purposes, then the firm is able to make use of the technological advantages appropriately.

The speediness changes we face globally, its complexity, and volatility (behavior change or social disturbances) might bring further twists that will impact the development of the ecosystems and the key role of marketing within them. They can not only be a limitation but also engage researchers for new directions. The authors recommend to explore the challenge in other consumption context in line with the digital transitions and to showcase the impact they have on consumer behavior.

Further researches could investigate the practical solutions to accompany our society into the society 5.0's level. It is very important in this time that we do not have time to hesitate, since literature defines it as disruptions. We also recommend to address the use of digital technologies in traditional context by managers and foreground the economic and social considerations.

In spite of the current developments in the food sector, it is proposed that research and studies are performed to investigate the feasibility of these new technologies for the small- and medium-scaled companies as they are the important enterprises in the food sector for many countries.

References

1. Kagermann, H., Lukas, W.D., Wahlster, W., Industrie 4.0: Mit dem internet der dinge auf dem weg zur 4. Industriellen revolution. *VDI Nachrichten*, 13, 1, 2–3, 2011.
2. Roblek, V., Mesko, M., Krapez, A., A complex view of industry 4.0. *SAGE Open*, 6, 2, 1–11, 2016.
3. Grewal, D., Noble, S.M., Roggeveen, A.L., Nordfalt, J., The future of in-store technology. *J. Acad. Market. Sci.*, 48, 1, 96–113, 2020.
4. Xu, L.D., Xu, E.L., Li, L., Industry 4.0: State of the art and future trends. *Int. J. Prod. Res.*, 56, 8, 2941–2962, 2018.
5. Kagermann, H., Change through digitization—Value creation in the age of industry 4.0, in: *Management of Permanent Change*, pp. 23–45, Springer Gabler, Wiesbaden, 2015.
6. Cova, B. and Cova, V., Tribal marketing: The tribalisation of society and its impact on the conduct of marketing. *European Journal of Marketing*, 36, 5/6, 595–620, 2002.
7. Camus, S., Proposition d'échelle de mesure de l'authenticité perçue d'un produit alimentaire. *Rech. Appl. en Mark. (French Edition)*, 19, 4, 39–63, 2004.
8. Lee, J. and Chung, L., Effects of perceived brand authenticity in health functional food consumers. *Br. Food J.*, 122, 2, 617–634, 2019.
9. Napoli, J., Dickinson, S., Beverland, M., Farrelly, F., Measuring consumer-based brand authenticity. *J. Bus. Res.*, 67, 6, 1090–1098, 2014.
10. Portal, S., Abratt, R., Bendixen, M., The role of brand authenticity in developing brand trust. *J. Strateg. Mark.*, 27, 8, 714–729, 2019.
11. Pelet, J.-E., Durrieu, F., Lick, E., Label design of wines sold online: Effects of perceived authenticity on purchase intentions. *J. Retail. Consum. Serv.*, 55, 102087, 2020.
12. Oltra-Mestre, M.J., Hargaden, V., Coughlan, P., Segura-García del Río, B., Innovation in the agri-food sector: Exploiting opportunities for industry 4.0. *Creativity Innov. Manage.*, 30, 1, 198–210, 2021.
13. VACEK, Jiří. On the road: From industry 4.0 to society 4.0, Trendy v Podn, pp. 43-49, 2017.
14. Horváth, D. and Szabo, R.Z., Driving forces and barriers of industry 4.0: Do multinational and small and medium-sized companies have equal opportunities? *Technol. Forecast. Soc. Change*, 146, 119–132, 2019.
15. Official website of the agricultural development agency: https://www.ada.gov.ma/ (Accessed March 19, 2021)
16. McKinsey Homepage, www.mckinsey.com, last accessed 2021/11/21.
17. Mumford, L., *The City in History: Its Origins, its Transformations, and its Prospects*, vol. 67, Houghton Mifflin Harcourt, Boston, MA, USA, 1961.
18. Firat, A.F. and Venkatesh, A., Liberatory postmodernism and the re-enchantment of consumption. *J. Consum. Res.*, 22, 3, 239-267, 1995.

19. Fukuyama, M., Society 5.0: Aiming for a new human-centered society. *Japan Spotlight*, 1, 47–50, 2018.
20. Fukuda, K., Science, technology and innovation ecosystem transformation toward society 5.0. *Int. J. Prod. Econ.*, 220, 107460, 2020.
21. Carayannis, E.G. and Morawska-Jancelewicz, J., The futures of Europe: Society 5.0 and industry 5.0 as driving forces of future universities. *J. Knowl. Econ.*, 1–27, 2022.
22. Hysa, B., Karasek, A., Zdonek, I., Social media usage by different generations as a tool for sustainable tourism marketing in society 5.0 idea. *Sustainability*, 13, 3, 1–27, 2021.
23. De Felice, F., Travaglioni, M., Petrillo, A., Innovation trajectories for a society 5.0. *Data*, 6, 11, 115, 2021.
24. Rojas, C.N., Peñafiel, G.A.A., Buitrago, D.F.L., Society 5.0: A Japanese concept for a superintelligent society. *Sustainability*, 13, 12, 6567, 2021.
25. Frederico, G.F., From supply chain 4.0 to supply chain 5.0: Findings from a systematic literature review and research directions. *Logistics*, 5, 3, 49, 2021.
26. Horváth, D. and Szabo, R.Z., Driving forces and barriers of Industry 4.0: Do multinational and small and medium-sized companies have equal opportunities? *Technol. Forecast. Soc. Change*, 146, 119–132, 2019.
27. Mittal, S., Khan Ahmad, M., Romero, D., Wuest, T., A critical review of smart manufacturing & industry 4.0 maturity models: Implications for small and medium-sized enterprises (SMEs). *J. Manuf. Syst.*, 49, 194–214, 2018.
28. Schmidt, R., Mohring, M., Harting, R.C., Reichstein, C., Neumaier, P., Jozinovic, P., Industry 4.0-potentials for creating smart products: Empirical research results, in: *International Conference on Business Information Systems*, Springer, pp. 16–27, 2015.
29. Kovacs, O., Az ipar 4.0 komplexitása–I. *Közgazdasági Szemle*, 64, 7-8, 823–854, 2017.
30. Arya, V., Verma, H., Sethi, D., Agarwal, R., Brand authenticity and brand attachment: How online communities built on social networking vehicles moderate the consumers' brand attachment, IIM Kozhikode Society & Management Review, 8, 2, 87-103, 2019.
31. Hernandez-Fernandez, A. and Lewis, M.C., Brand authenticity leads to perceived value and brand trust. *Eur. J. Manage. Bus. Econ.*, 28, 3, 222–238, 2019.
32. Lee, J. and Chung, L., Effects of perceived brand authenticity in health functional food consumers. *Br. Food J.*, 122, 2, 617–634, 2019.
33. Uchiyama, K. and Shiroishi, Y., Society 5.0: For human security and well-being. *IEEE Comput. Soc.*, 51, 7, 91–95, 2018.
34. Almeida, F., Concept and dimensions of web 4.0. *Int. J. Comput. Technol.*, 16, 7, 7039–7046, 2017.
35. Grewal, D., Noble, S.M., Roggeveen, A.L. *et al.*, The future of in-store technology. *J. Acad. Mark. Sci.*, 48, 96–113, 2020.

36. Verhoef, P.C., Kannan, P.K., Inman, J.J., From multi-channel retailing to omni-channel retailing: Introduction to the special issue on multi-channel retailing. *J. Retailing*, 91, 2, 174–181, 2015.
37. Aujourdhui Homepage, aujourdhui.ma/culture/reseaux-sociaux-22-millions-dutilisateurs-au-maroc, last accessed 2021/09/21.
38. Hair, J., Risher, J., Sarstedt, M., Ringle, C., When to use and how to report the results of PLS-SEM. *Eur. Bus. Rev.*, 31, 1, 2–24, 2019.
39. Sarstedt, M., Hair Jr., J., Cheah, J., Becker, J., Ringle, C., How to specify, estimate, and validate higher-order constructs in PLS-SEM. *Australas. Mark. J.*, 27, 3, 197–211, 2019.
40. Hasnan, N.Z.N. and Yusoff, Y.M., Short review: Application areas of industry 4.0 technologies in food processing sector, in: *2018 IEEE Student Conference on Research and Development (SCOReD)*, Doctoral dissertation, IEEE, pp. 1–6, 2018.
41. Benamour, Y., Confiance interpersonnelle et confiance institutionnelle dans la relation client-entreprise de service: Une application au secteur bancaire français, Doctoral dissertation, Paris, 9, 2000.
42. Price, L. and Arnould, E., Commercial friendships: Service provider–client relationships in context. *J. Mark. Manage.*, 63, 4, 38–56, 1999.

Appendix

Table 12.4 Measures for variables.

Construct	Scale of measurement (7 points likert)
AQ (Authenticity quest)	Items adapted from [9-7]
RAc/DAc (Trust in Real/Digital)	Items adapted from [41]
RBI/DBI (Real/Digital Behavior)	Items adapted from [42]

13

IoT-Based Smart City Applications: Infrastructure, Research and Development

Tarush Gupta, Princy Randhawa and Nikhil Vivek Shrivas

Department of Mechatronics Engineering, Manipal University Jaipur, Jaipur, India

Abstract
Internet of things is a gateway to link several objects aiming to communicate with each other and parallelly easing the human manual work such that the same work does not require the amount of energy it usually does, and the muscle work is done by the machine with a greater accuracy than what is done by the human but the new problem the world faces is the problem of sustainability, which can be devastating. What we fail to understand is that sustainability is not just the careful use of natural resources, it includes social, economic, and environmental factors. While developing better cities and better living conditions, we often forget the role of people and the number of resources it can drain. This chapter aims to provide surveys and analysis of the smart city projects and how we can sustain them and balance the planet, people, and profits. Also, suggesting new technological implementations that are not just appealing to us but solve actual problems in developing countries like India.

***Keywords*:** Smart city, Internet of Things, digital city, sustainability, smart planet, smart devices

13.1 Introduction

When we talk about the term "smart," we automatically refer to something that can complete an action by itself with minimum human interaction. In recent years, we humans have focused on automation, which has brought

**Corresponding author*: princyrandhawa23@gmail.com

Loveleen Gaur, Vernika Agarwal and Prasenjit Chatterjee (eds.) Decision Support Systems for Smart City Applications, (219–234) © 2023 Scrivener Publishing LLC

up several fields and one of them is what we today knew as IoT or the Internet of Things that enables a machine to act on its own in a similar way the human body does. Talking about how IoT and humans are interrelated. Humans have five senses, namely vision, hearing, smell, taste, and touch, our body uses these senses to gather information from our surroundings, and these data are then converted into electrical signals that through millions of neural nerves reaches the brain, which has the capability of processing this information on its own and reacting to it according to the conditions, it decides the reaction to the environment and accordingly, it signals the other body parts to react to it through hands, legs, mouth, etc. Similar is how IoT devices works, there are three major parts of an IoT device/system, the sensors, the microcontroller, the actuators [1–3]. Sensors are the devices that gather information from the environment and convert them into electrical signals and transmits those signals to the microcontroller that is programmable and once that information is processed and matched with the criteria set, the microcontroller then transmits electrical signals to the actuators where the electrical energy is converted into mechanical energy or light or sound, etc. as shown in Figure 13.1.

Figure 13.1 represents the architecture of how an IoT device works. It basically consists of three individual elements that are enabled with inter-device communication to make a successful IoT device. The sensors gather information from the environment and communicate with the microcontroller (brain of the system), the microcontroller then acts on the information given by the sensors by communicating with actuating peripherals that enable an action like producing a sound, turning on a motor, lighting up an LED, etc.

Figure 13.2 shows how the IoT incorporated into smart cities development has gained over the years. The remarkable research has been explored in IoT-based smart city systems and projects. The project was first mentioned in the year 2009 in the strategic energy technology plan which aimed at a more efficient running of cities and towns with the help of installation of smart systems enough to sustain a city needs on its own, without excessive labor, repeated investments, and regular significant modifications. Arguably, the first digital city was worked on in Amsterdam in 1984, which made to increase communications, but the first actual foundations of the

Figure 13.1 Architecture of an IoT system.

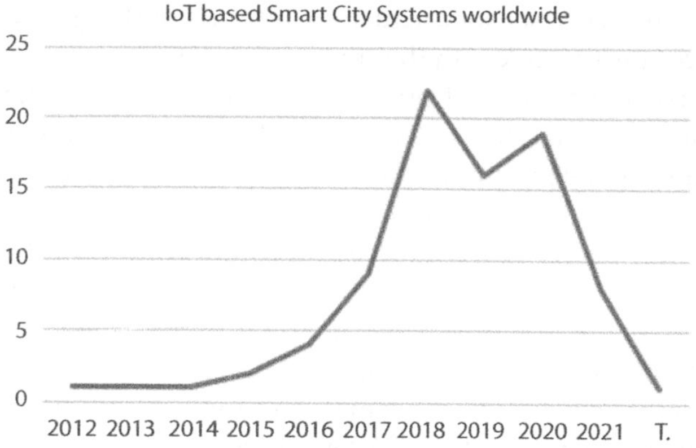

Figure 13.2 IoT-based smart system over the years.

smart city projects were laid in 2009 not only in Amsterdam but also in several other cities and countries.

The initial phases of the smart cities were based on optimization of energy, better connectivity between people and the networks, and open data. These building blocks enable each city to advance as solving these basic problems opens room for development without the requirement to fix problems like basic connectivity or high power consumption.

13.2 Different Phases of Development

13.2.1 Investments, Research, and Planning

The initial phases of every developing smart city in the world right now began in 2009 with countries like Amsterdam, Madrid, Japan, etc. In 2009, the SETP mentioned the term Smart City but before this serval countries had started raising money, to begin with, the smart city project but for Amsterdam, the foundations had been laid in 1984 when projects like digital cities came up increasing the use of the internet and improving communications. The digital city project started in Amsterdam, and it laid down the foundation of the smart city systems in Amsterdam right now. It aimed at Internet connectivity and awareness about the Internet how it can be beneficial to businesses, jobs, shops, services, etc. Once the people were educated, the city of Amsterdam became digital where everything in the city is accessible through the Internet. In 2009, when the term smart city was introduced, companies like Cisco, IBM, invested millions of dollars on smart city projects, sensors, actuators, Internet of things, etc. The main aim for Cisco was to work with the smart cities developing in Amsterdam, San Francisco, and Seoul and prove the effectiveness and importance of these technologies and later on commercialize the products and services developed during the program, while IBM invested in research over sensor technology, device communication, etc. and integrate it with urban cities. Along with them, "The USA also passed the American Recovery and Reinvestment Act (ARRA) provided funding for US Smart Grid Projects [4, 5]. The two major grants were the Smart Grid Investment Grant (SGIG) and Smart Grid Demonstration Program (SGDP). The SGIG deploys systems on current smart grids, machines, and tools, techniques to improve grid performance whereas the SGDP explores and researches smart grids and energy storage systems for future performance and application analysis" [6–8].

In 2010, Japan named its first Smart City Demonstration Project Yokohama partnered with Frankfurt as the first Smart City Demonstration Project. YSCP was a program that included 34 Japanese companies, with its funding issued by the Ministry of Economy, Trade, and Industry. In addition to obtaining the cooperation of Yokohama's small- and medium-sized builders, etc., concerning HEMS introduction, during the project's demonstration phase, efforts were made to attract the cooperation of a wide variety of stakeholders, including numbers of participating citizens [9, 10].

Between 2009 and 2010, several other countries started investing in smart city systems and billions of dollars were spent on various projects based on smart city research and development, some of them directly investing in smart city projects and demonstrations while the rest devoting research on different individual systems which in the future would lay down the foundations of the smart cities. Several countries and companies invested their time and money in research in sensors, controllers, actuators and use that information to develop smart technologies with applications in the smart city projects worldwide.

Shortly after the initial investments and research work, several countries and companies were able to take the first steps toward the smart city projects by the mid of 2011. While the projects like SGIG were in their guidance and deployment phases, Cisco's and IBM's early investments were fruitful [11].

13.2.2 Execution and Deployment

In 2011, IBM was able to deploy some of its research to use in the capital of Brazil, Rio de Janeiro. Rio's heavy rainfalls are very uncertain, and they cause heavy landslides all of the capital, IBM was able to deploy weather sensors in different parts of the cities, these weather sensors were able to predict the rainfalls 24 hours prior, and since the emergency services were reported to the response center, the response to such conditions was given well in time. This system that was developed at the end of 2010 proved its worth and importance and now is available commercially. The next project IBM took up was in Glasgow, where fuel is scarce. In 2011 IBM used its technology to sense heat dissipation from homes and nearby industries and use it to warm homes and provide heat to the people. Cisco's advancements focused on traditional city systems like improvement of rail and station monitoring systems with the partnership with the Metropolitan Transit Authority of New York and similar other technologies. In 2011, IBM granted 24 cities with a special smart city grant which provided the local municipalities with expertise and company technology to build toward a smarter future.

13.2.3 Globalization of IoT-Based Technologies

In 2011, came up the first Smart City Expo World Congress in Barcelona where more than 6000 people joined in from different countries all around the globe to discuss the smart city projects. This expo aimed at how a smart city can be made livable and sustainable on its own. These events gathered interest from all around the world, which gave exposure to this project with people and countries that were unaware of it.

In 2012, Barcelona used responsive technologies in urban systems like public transportation, parking, street lighting, and waste management system. They used their 30-year-old fiber optic technology that was spread throughout the city to integrate smart systems. These fiber optics are now connected to almost 90% of homes in Barcelona which is now the backbone of the strong communication within the whole city. This has further helped the integration of management systems that control and automate the public transit, waste management system, etc.

This phase of early outcomes began a new attraction toward IoT-based smart cities as they not only proved their worth and effectiveness but also were cost-effective and lifesaving and several other countries came up with new projects and demonstrations in the Smart City Field.

13.2.4 Pilot Projects and City Designs

At the end of 2012, China announced its first smart city projects which included 90 cities, towns, and districts. These cities, towns, and districts had distinctive problems and potential of better living and economic conditions which could be resolved by the integration of IoT. China started with five pilot cities from test runs of different systems suiting to the needs of each city and condition.

The first city among the five was Guilin where China introduced the Smart Tourism Project, under the govt. used serval systems like cloud computing, IoT, Intelligent applications, etc. to build a smarter application system for tourists as well as the tourism department. "This system was said to be 'One platform — four systems — two demonstration projects — series of eight tourism products'. This one platform comprised of the four systems namely service system, management system, marketing system, and tourism enterprising system, two demonstration projects one characteristic small town and one low carbon characteristic small town, and finally eight tourism products, namely water sighting, leisure vacation, historical and cultural tourism, red tourism, ecological village tourism, outdoor sports tourism, and romantic wedding tourism."

The second city on the list was Yun long, which from the beginning was planned as an eco-friendly low-carbon town; it was divided into three parts, cultural park area, tourism area, and a new industrial area. To enhance infrastructure command operations, ability to respond to emergencies each part was further divided into 6 parts: low carbon energy, green traffic, comprehensive utilization of water resources, ecological and environmental protection, and low carbon industries and urban management. This city was specially made to focus on smart utilization of water resources as it had a constant supply from nearby streams [12, 13].

The next was the district of Panyu which was one of the fastest-growing districts in terms of population, and hence, the main focus to integrate smart city systems was to improve the livelihood of the people and provide better information services. In this district, they started four projects, namely the file management system, livelihood service cards, service points, and livelihood services. They particularly promoted the cloud system that was integrated with the smart service points and the smart app to push the construction of smart informative systems, providing the public with several services.

The fourth city in list Yangling demonstration zone was agricultural land and the people and the government believed that the smart city systems could be integrated within the zone to provide better farming facilities and consume interaction. This zone was based on the belief that modern technologies were important in the field of agriculture to promote better production, food safety, and quality management. A system was developed where each produced product was given an "identity card" in the form of codes which the consumer could scan to find the details, like sowing time, fertilization time, fertilizer used, etc. [14–16].

Finally, the city Lecong which focused on the industrial integration of IoT, in town was able to focus on automation of several industries-based works and has laid an example from different other cities and countries how to move forward with industrial automation and integration of IoT.

These five pilot cities showed a huge amount of potential within the next few months and in 2014 China announced the second batch the smart cities consisting of 103 pilot cities, this announcement made it clear that all the systems deployed by China in the first five cities proved their efficiency, effectiveness, and potential.

In 2013, the mayor of London introduced the first Smart London Board which aimed at shaping the digital city technology in London. This board aimed at providing digital-driven improvements to London and delivering data. They also aimed at strategic partnerships to bring new resources to their smart city systems, and they also introduced the ambassadors to smart cities, which was important for making intercommunity bonds.

By 2013 and 2014, previous projects had reached good stages, for example, the SGIG and SGDP were already in their analysis and development stage, and they were learning from their errors and constantly improving their systems.

Vienna City council in 2014 adopted a Smart city Wein Framework Strategy, which is set Vienna on a path to sustainable smart city development shortly. This strategy involved experts from the different departments with a shared vision of sustainable development going hand in hand with worldwide technological developments. Vienna managed to stick to its basic values and kept the quality of life to utmost importance keeping it traditionally unchanged, while they aimed to equip the city with smart systems that new technological and informative capabilities, climate action and resource development, better opportunities for better involvement of people to inspire life and technology [17–20].

In 2015, China launched another pilot run of 84 cities now bringing up the total number of smart cities in China to 277 in all, out of which more than half of the cities were running successfully and constantly developing new technology [21].

With a budget of over 14 billion US dollars, in 2015, India launched its first 100 smart city projects, India a vast terrain had a strategy to develop area-wise smart systems. This area-based development aimed at improving economic viability, effectiveness, and most important of all the main purpose of this was to make cities and towns more livable, and this strategy not just covered cities and towns, it included several village areas, as well as slums. The strategy was divided into four parts, namely retrofitting, redevelopment, greenfield development, and a pan-city initiative [22].

Retrofitting: In this model, the aim is to make currently present areas more livable by demolishing or changing the infrastructure. This model recognizes 500 acres of land recognized and consulted by the citizens as well as the government to develop smart city systems within the existing infrastructure. This model will not require a lot of time as the intensive development of smart systems will be done in the existing infrastructure itself [23, 24].

Redevelopment: Using this strategy India aimed to replace the existing built-up environment and enable the co-creation of a new layout with the enhanced infrastructure using mixed land use and increased density. This creates the possibility of over 50 acres of land to be identified by the negotiation with the citizens and the Urban Local Bodies. Saifee Burhani Uplifting Project in Mumbai and the redevelopment of East Kidwai Nagar in New Delhi bring undertaken are examples of redevelopment.

Greenfield Development: With this model Indian planned on deploying smart city systems and tools to address the growing population of the country. This strategy is aimed at using the vacant areas all over India using innovative planning and plan financing to make livable housing provide it to the poor at affordable prices. One such project is the GIFT city in Gujarat [25].

Pan-City Development: This creates a possible application of a few Smart Solutions to the existing city infrastructure. This model aims to integrate smart systems with the existing technology. For example, applying smart systems to the transportation sector like intelligent traffic systems, smart routes are all aimed toward the improvement of the quality of life of the people [26].

Since 2015 India has constantly been deploying smart city systems all over the country some of the few examples are the smart city center in Surat monitoring communication, planning, and organizing throughout the city, Smart street benches throughout the city of Coimbatore with free Wi-Fi all over the city, The All Ability Children Park in Visakhapatnam aiming to seek the balance of the differently-abled children without segregating them from the balance of the community, Intelligent Transportation Management in Pune, Public Bicycle sharing in Bhopal, etc.

After the Smart City technology proved its potential and importance in early years, throughout the world several companies and countries organized Smart City or Smart Technology Challenges to promote the development of smart systems and smart cities. A similar challenge was given out by the US Department of Transportation in 2016 for smart transportation systems like self-driving cars or connected vehicles. Columbus city from Ohio won this price of $50m. Columbus connect the transportation network, integrated data exchange enhanced the human services, developed electric vehicle infrastructure in residential districts, commercial districts, downtown districts, and the logistics districts [27].

While the technology throughout the world was at its full pace, it was time that the backbone of this system had the potential to grow. In 2017 UK started a trial program and testbeds for the 5G network, in 2016 UK announced its plan to invest in the 5G research which was over a billion euros. This program looked to utilize the areas where the UK has a competitive advantage in fields such as scientific research, engineering talent, and their rich variety of technology business. The key objectives of this program were to accelerate the deployment of 5G networks and be advantageous to the UK, maximize efficiency and profits using 5G and create new opportunities for the UK.

While the smart city revolution was taking over the world, Hong Kong in late 2017 released a blueprint setting out 76 initiatives under 6 categories namely Smart Mobility, Smart Living, Smart Environment, Smart People, Smart Government, and Smart Economy. This was worked on for the past 3 years and in 2020 Hong Kong released its Blueprint 2.0 that included 60 new initiatives in the original blueprint some of them are:

- Smart Mobility: Develop Traffic data management and efficiency; Smart traffic system to develop and research application of vehicle-related I&T
- Smart Living: Use the "iAM Smart" platform to streamline the Transport Department's licensing services; Explore the use of telehealth, video conferencing, and remote consultation in Hong Kong.
- Smart Environment: Launch the "Smart Toilet" pilot program and explore the related technologies; Improve pest control using technologies like IoT
- Smart People: Implement the IT innovative Lab in Secondary schools; Continue to implement STEM Internship Scheme
- Smart Government: Develop electronic hub for submission for processing building plans; Facilitate online licensing, government service applications, etc.
- Smart Economy: Develop online platform for dispute resolution and deal-making services; Develop eMPF Platform by the Mandatory Provident Fund Schemes Authority [28].

13.3 Current Scenario

Since 2018, the pace of technological developments throughout the world has increased massively from dense countries like India deploying smart systems in several of its cities to China's 277 pilot smart city testing, from small countries rising the economy by the deployment of these smart city systems, to giant projects like that of The USA all have proved the importance and the high potential of growth using these types of systems. These analytics and results gathered more and more attention of people throughout the world and several other gigantic companies like Google, Amazon, Ford, etc. dug their hands deeper into automation and IoT-based smart systems leading the way to new technologies for a better future [34].

In 2018 Toronto partnered with Google offshoot Sidewalk Labs to develop Toronto's water-edged regions with smart features like snow-melting roadways, underground delivery systems, data collection sensors. Although the project was shut off in May 2020 but was still somehow proved an effective system [29].

The London Smart Board introduced in 2013 updated its roadmap in the year 2018 to make London the smartest city in the world. They changed strategies to have a better, more flexible digital plan of action. They aimed at increased efficiency in form of 5 different missions. These missions were:

- More User-Designed Services
- Striking new deal for city data
- World-Class connectivity and smarter streets
- Enhanced digital and leadership skills
- Improved city-wide collaboration [30].

13.3.1 Achievements and Milestones

While each country was thriving hard to excel in the Smart City development the IESE ranked the top three cities which were: New York, Paris, and London. IESE Cities in Motion is a research initiative that brings together a global network of experts in smart cities and specialized private companies with local governments from around the world. The aim was to promote technologies and new changes at the local level and develop them into valuable ideas and innovative tools for sustainable and smarter cities. The 4 main factors the IESE analyses are sustainability, ecosystem, innovative activities, and fairness among citizens and connected territory.

In the year 2018, Singapore won the Smart City Expo World Congress award for being a "meaningful" smart city. It was praised for its wide array of government-developed solutions, from dynamic public bus routing algorithms to real-time parent-teacher portals and predictive analytics for water pipe leaks. Some other awards awarded in the Expo were:

- Innovative Idea: This award was rewarded to Cape Town in Africa, where housing is a big problem as there are low-wage earners in the city. A mobile application was developed which connected the people, investors, construction companies to provide people with houses.
- Digital Transformation: This award was given to the Government of Gaoqing for the deployment of smart ICT systems.
- Inclusive and Sharing City: This award was given to The Hague city in the Netherlands for improving and developing smart health conditions of aging citizens using their feedbacks.
- Governance and Finance: This award was given to the city of Msheireb Downtown Doha in Qatar for a citizen-focused environment integrated with smart systems.
- Mobility: This award was granted to Atlanta City, USA for its smart traffic system model which has improved roadways, public safety, mobility, and a better environment.
- Urban Environment: Shanghai in China won this award for managing health and the environment while going hand in hand with more industries. Their smart buildings have made this possible and shown huge potential [31].

In 2019 Ford declared that they would introduce C-V2X technology in their cars which will be functioning on the 5G technology. This tech plans to increase road safety and traffic management. The Car will be able to communicate with the user by "talking and listening." These cars can link with traffic management systems like traffic lights, smart roadways, routes, etc. They are also developed keeping human safety in mind where users can connect using a mobile application and the car will be able to read it and respond accordingly. Similarly, other motor companies started developing smart auto systems example KIA motors is developing a smart car environment system based on AI that changes the environment of the car by the emotions of the driver.

In 2019, the G20 summit lead the way for the world and began to implement smart city norms for the whole world. A Global Smart Alliance was made that focuses on how smart city technology is used in public spaces and it focuses on the values of core principles like transparency, security, and privacy. This was discussed as the main agenda in this G20 summit where a global framework was being established which did not exist before. This effort was put in to establish trust and security among people to prevent any misuse of data collected by smart city technology. After this summit, The Internet of Things, Robotics, and Smart city team of the Forum's Centre for the Fourth Industrial Revolution has taken lead to ensure accountability of all the members of the Global Smart Alliance.

13.3.2 IoT in Smart Cities

IoT or the Internet of things was one of the most important technologies that have been used in the development of Smart Cities from the very beginning. Since smart cities are based on the concept of gathering data and information and act "smartly" on the given information, IOT lays the very foundation of Smart city systems, IoT enables the very first step to Smart City development, as it enables the data collection with the use of sensors, transducers, etc. Sensors and transducers are devices that convert one form of energy into the other, e.g. A sound sensor converts sound energy to electrical energy, which is then used as a transfer of data to another device using wireless/wired electrical connections.

In areas like Rio where the heavy rainfalls were uncertain, the system that was developed to overcome this problem was based on data collected from the environment using weather sensors which collected data and converted it to electrical energy, and then the concepts of deep and machine learning would make statistical predictions, the electrical signals of which are again communicated to the response systems and the action would be taken. IoT enables any device with a system capable to collect and share information device to device without any human involvement.

IoT not only enables data transmission and collection it can enable a device to act on its own, in places like Yanling, China where smart agricultural systems were established, the data collected from the sensors was not only used to provide the consumer with the complete details of the crop but were also used to manage the quality of any degrading crop by automatic water sprinklers, fertilization management, climate management systems, etc. IoT establishes the base and the central control system of any automated device, as it can enable internetwork communications within several devices at once [32–34].

13.4 Conclusion and Future Work

While the present has a great pace of development, a lot of countries and companies have already started building for a better future. With the revolutionary Smart City technology, every nation was trying to build a new kind of life with better quality for its citizens. It is expected that up to 70% of the worlds' population will live in cities, and hence, the cities need to be more livable, secure, and smart. Increased job opportunities, climate control and response system, environment sustainability, etc. will be very important in the upcoming years and can only be achieved through development in smart technologies all over the world. Several counties are in the process of funding more futuristic projects of smart city systems, and countries, like Vietnam, have already invested billions of dollars in their smart city projects. The Covid-19 pandemic slowed down a lot of research and development opportunities for countries around the world but now they are getting back on track and soon will be on the pace of development stronger and faster than before.

Newer researches in the areas of industry 4.0 and 5.0 are aiming to automate several devices in order to increase the efficiency, with better sensing technologies to predict and handle situations that are unfavorable for human survival. Such developments are taking one step closer to the formation of an automated environment creating safer, more efficient, and comfortable living conditions for human survival.

References

1. Kumar, S., Sahoo, S., Mahapatra, A., Swain, A.K., Mahapatra, K.K., Security enhancements to the system on chip devices for IoT perception layer, in: *2017 IEEE International Symposium on Nanoelectronic and Information Systems (INIS)*, IEEE, pp. 151–156, December 2017.
2. Ande, R., Adebisi, B., Hammoudeh, M., Saleem, J., Internet of things: Evolution and technologies from a security perspective. *Sustain. Cities Soc.*, 54, 101728, 2020.
3. Monzon, A., Smart cities concept and challenges: Bases for the assessment of smart city projects. *2015 International Conference on Smart Cities and Green ICT Systems (SMARTGREENS)*, IEEE, 2015.
4. Borlase, S. (Ed.), Smart Grids: Infrastructure, technology, and solutions, CRC Press, Boca Raton, 2017.
5. Liu, X., Marnay, C., Feng, W., Zhou, N., Karali, N., A review of the American recovery and reinvestment Act Smart Grid Projects and their implications for China, Lawrence Berkeley National Laboratory, China, 2017.

6. Biviji, M.A., Martini, L., Nigris, M.D., Kang, D.J., Ton, D., Global survey of smart grid activities, in: *Smart Grid Handbook*, pp. 1–22, 2016.
7. Karali, N., Marnay, C., Yan, T., He, G., Yinger, R., Mauzey, J., Zhu, H., Towards uniform benefit-cost analysis for Smart Grid Projects: An example using the Smart Grid computational tool, Lawrence Berkeley National Laboratory, USA, 2015.
8. Perboli, G., De Marco, A., Perfetti, F., Marone, M., A new taxonomy of smart city projects. *Transp. Res. Proc.*, 3, 470–478, 2014.
9. Van de Graaf, T., A new world: The geopolitics of the energy transformation IRENA – International Renewable Energy Agency, UAE, 2019.
10. Su, K., Li, J., Fu, H., Smart city and the applications. *2011 International Conference on Electronics, Communications and Control (ICECC)*, Ningbo, China, pp. 1028–1031, 2011.
11. Kumar, T.V., Smart metropolitan regional development: Economic and spatial design strategies, in: *Smart Metropolitan Regional Development*, pp. 3–97, Springer, Singapore, 2019.
12. Yigitcanlar, T., Technology and the city: Systems, applications, and implications, London, 2016.
13. Hassan, Q.F. (Ed.), Internet of Things A to Z: Technologies and applications, John Wiley & Sons, 2018.
14. Zhang, Y., Interpretation of smart planet and smart city. *China Inf. Times*, 10, 38–41, 2010.
15. Paskaleva, K.A., Enabling the smart city: The progress of city E-governance in Europe. *IJIRD*, 4, 405–422, 2009.
16. Perboli, G., De Marco, A., Perfetti, F., Marone, M., A new taxonomy of smart city projects. *Transp. Res. Proc.*, 3, 470–478, 2014.
17. Kim, S.Y., *Design as a Strategic Asset in Visual City Branding*, Doctoral Dissertation, Lancaster University, 2017.
18. Benjelloun, A., Crainic, T.G., Bigras, Y., Towards a taxonomy of City logistics projects. *Proc. Soc. Behav. Sci.*, 2, 3, 6217–6228, 2010.
19. Auci, S. and Mundula, L., Smart cities and a stochastic frontier analysis: A comparison among European Cities, *SSRN Electronic Journal*, Available at SSRN 2150839, 2012.
20. Dameri, R.P., Searching for smart city definition: A comprehensive proposal. *Int. J. Comput. Technol.*, 11, 5, 2544–2551, 2013.
21. Mora, L. and Deakin, M., *Untangling Smart Cities: From Utopian Dreams to Innovation Systems for a Technology-Enabled Urban Sustainability*, Elsevier, 2019.
22. Gharaibeh, A., Salahuddin, M.A., Hussini, S.J., Khreishah, A., Khalil, I., Guizani, M., Al-Fuqaha, A., Smart cities: A survey on data management, security, and enabling technologies. *IEEE Commun. Surv. Tut.*, 19, 4, 2456–2501, 2017.

23. Ahas, R., Mooses, V., Kamenjuk, P., Tamm, R., Retrofitting soviet-era apartment buildings with 'Smart City' features: The H2020 smarten city project in Tartu, Estonia, in: *Housing Estates in the Baltic Countries*, p. 357, 2019.
24. Van Winden, W., Oskam, I., van den Buuse, D., Schrama, W., van Dijck, E.J., Organising Smart City Projects: Lessons from Amsterdam, Amsterdam, Hogeschool van Amsterdam, 2016.
25. Shcherbina, E. and Gorbenkova, E., Smart city technologies for sustainable rural development, in: *IOP Conference Series: Materials Science and Engineering*, vol. 365, IOP Publishing, p. 022039, June 2018.
26. Gupta, K., Zhang, W., Hall, R.P., Risk priorities and their co-occurrences in smart city project implementation: Evidence from India's smart cities mission (SCM). *Environ. Plan. B Urban Anal. City Sci.*, 2399808320907607, 4, 2020.
27. Shladover, S.E. and Bishop, R., *Road Transport Automation as a Public–Private Enterprise*, 2015.
28. Lam, Patrick T.I., and Yang, W., Factors influencing the consideration of Public-Private Partnerships (PPP) for smart city projects: Evidence from Hong Kong, 102606, 2020..
29. Laura, M., Liu, R., Valois, M.-F., Xu, J., Weichenthal, S., and Hatzopoulou, M., Development and comparison of air pollution exposure surfaces derived from on-road mobile monitoring and short-term stationary sidewalk measurements. *Environmental Science & Technology*, 52, 6, 3512-3519, 2018.
30. Greater London Authority, City Hall, The Queen's Walk, London SE1 2AA, June 2018.
31. Wood, D.M., and Mackinnon, D., Partial platforms and oligoptic surveillance in the smart city. *Surveillance & Society,* 17, no. 1/2: 176-182, 2019.
32. Hamidreza, A., Hosseinnezhad, V., Loia, V., Tommasetti, A., Troisi, Q., Shafiekhah, M., and Siano, P., Iot based smart cities: A survey. In *2016 IEEE 16th international conference on environment and electrical engineering (EEEIC)*, IEEE, pp. 1-6, 2016.
33. Scuotto, V., Ferraris, A. and Bresciani, S. (2016), Internet of Things: Applications and challenges in smart cities: A case study of IBM smart city projects, *Business Process Management Journal*, 22, 2, 357-367. https://doi.org/10.1108/BPMJ-05-2015-0074.
34. Benjelloun, A., Crainic, T.G., Bigras, Y., Towards a taxonomy of city logistics projects. *Proc. Soc. Behav. Sci.*, 2, 3, 6217–6228, 2010.

Index

Also of Interest

Check out these published and forthcoming titles in "Sustainable Computing and Optimization" series from Scrivener Publishing

Optimization Techniques in Engineering
Advances and Applications
Edited by Anita Khosla, Prasenjit Chatterjee, Ikbal Ali and Dheeraj Joshi
Forthcoming 2022. ISBN 978-1-119-90627-8

Computational Intelligence in Sustainable Reliability Engineering
Edited by S. C. Malik, Deepak Sinwar, Ashish Kumar, S. R. Gadde, Prasenjit Chatterjee and Bui Thanh Hung
Published 2022. ISBN 978-1-119-86501-8

Decision Support Systems for Smart City Applications
Edited by: Loveleen Gaur, Vernika Agarwal and Prasenjit Chatterjee
Published 2022. ISBN 978-1-119-89643-2

Fuzzy Computing in Data Science
Applications and Challenges
Editors: Sachi Nandan Mohanty, Prasenjit Chatterjee and Bui Thanh Hung
Published 2022. ISBN: 978-1-119-86492-9

Handbook of Intelligent Computing and Optimization for Sustainable Development
Edited by: Mukhdeep Singh Manshahia, Valeriy Kharchenko, Elias Munapo, J. Joshua Thomas and Pandian Vasant
Published 2022. ISBN 978-1-119-79182-9

Machine Learning Algorithms and Applications
Edited by Mettu Srinivas, G. Sucharitha and Anjanna Matta
Published 2021. ISBN 978-1-119-76885-2

Smart Sustainable Intelligent Systems
Edited by Namita Gupta, Prasenjit Chatterjee and Tanupriya Choudhury
Published 2021. ISBN 978-1-119-75058-1

www.scrivenerpublishing.com

Printed and bound by CPI Group (UK) Ltd, Croydon, CR0 4YY
27/10/2024
14580133-0001